FECAL MATTERS IN EARLY MODERN LITERATURE AND ART

To Erasmus, patron saint of intestinal disorders
Ora pro nobis.

He works with human excrement – what is rejected, what is accounted of no worth to mankind – and in it I suppose he hopes to discover something that is of worth.
– Robertson Davies, *The Rebel Angels*

Fecal Matters in Early Modern Literature and Art

Studies in Scatology

Edited by

Jeff Persels

Russell Ganim

Studies in European Cultural Transition

Volume Twenty One

General Editors: Martin Stannard and Greg Walker

ASHGATE

Published by
Ashgate Publishing Limited
Gower House
Croft Road
Aldershot
Hampshire GU11 3HR
England

Ashgate Publishing Company
Suite 420
101 Cherry Street
Burlington, VT 05401-4405
USA

Ashgate website: http://www.ashgate.com

British Library Cataloguing in Publication Data
Fecal matters in early modern literature and art : studies
in scatology. – (Studies in European cultural transition)
1. French literature – 16th century – History and criticism
2. French literature – 17th century – History and criticism
3. English literature – Early modern, 1500-1700 – History and criticism 4. German literature – Early modern, 1500-1700 – History and criticism 5. Scatology in literature 6. Scatology in art
I. Persels, Jeff II. Ganim, Russell
809.9'3353'09031

Library of Congress Cataloging-in-Publication Data
Fecal matters in early modern literature and art : studies in scatology / edited by Jeff Persels and Russell Ganim.
p. cm. – (Studies in European cultural transition)
Includes bibliographical references (p.) and index.
ISBN 0-7546-4116-3 (alk. paper)
1. Scatology in literature. 2. Scatology in art. I. Persels, Jeff. II. Ganim, Russell. III. Series.

NX650.S28F43 2004
700'.453—dc22

2003023570

ISBN 0 7546 4116 3

Printed and bound in Great Britain by MPG Books Ltd, Bodmin, Cornwall

Contents

List of Figures

General Editors' Preface

The European dimension of research in the humanities has come into sharp focus over recent years, producing scholarship which ranges across disciplines and national boundaries. Until now there has been no major channel for such work. This series aims to provide one, and to unite the fields of cultural studies and traditional scholarship. It will publish the most exciting new writing in areas such as European history and literature, art history, archaeology, language and translation studies, political, cultural and gay studies, music, psychology, sociology and philosophy. The emphasis will be explicitly European and interdisciplinary, concentrating attention on the relativity of cultural perspectives, with a particular interest in issues of cultural transition.

Martin Stannard
Greg Walker
University of Leicester

Acknowledgments

Versions of the articles in this volume, with the exception of Josef Schmidt's, were originally presented over two years at annual meetings of the Sixteenth Century Society, in sessions specifically devoted to scatological rhetoric. The editors wish expressly to thank the following: their fellow scatologs, for their contributions to this volume and endurance in seeing it through to completion; Erika Gaffney, for being an early and resilient champion of this project; Dora Polachek, for her valued contributions to the above-mentioned SCSC sessions; Mary McKinley and George Hoffmann, for their stylistic good sense; Peter Bleed for his advice on anthropological references; Joy Suder and Jason Cruise for their technical assistance; Karen James and the helpful staffs of the libraries and archives who provided information regarding or prints of the actual images reproduced in the volume; and finally, the mature patience and loving tolerance of Lisa Roberts, Madeleine Roberts-Ganim and Brigitte Guillemin Persels, as well as of the friends and colleagues of two men who still delight in playing in the dirt and who have found there great spiritual, if not material, riches.

Notes on Contributors

Barbara C. Bowen taught for 25 years at the University of Illinois in Urbana–Champaign and for 15 years at Vanderbilt University, before retiring in 2002. She has published or edited 10 books and 60 articles on Renaissance literary topics from Rabelais and Montaigne to art history and German Humanism, but most of her work relates in some way to Renaissance humor. After a dissertation on French farce from 1450 to 1550, she has explored wit and the comic in Rabelais, Tahureau, Beroalde de Verville, sixteenth-century emblems, and *facetiae* (jokes/jests) in a variety of languages.

Glenn Ehrstine is Associate Professor of German at the University of Iowa. He is the author of *Theater, Culture, and Community in Reformation Bern, 1523–1555* (Brill, 2002), winner of the 2003 David Bevington Award of the medieval and Renaissance Drama Society. His current research concerns devotional imagery and the representation of *figurae* in late medieval theater.

Russell Ganim is Associate Professor of French and Chair of the Department of Modern Languages at the University of Nebraska–Lincoln. He has published on the French Baroque Lyric (*Renaissance Resoncance: Lyric Modality in La Ceppède's Théorèmes*, Rodopi 1998), and co-edited with Anne L. Birberick *The Shape of Change: Essays in Early Modern Literature in Honor of David Lee Rubin* (Rodopi 2002). He is also co-editor of *EMF: Studies in Early Modern France* and its monograph series, *EMF Critiques*.

Geoffrey R. Hope is Professor of French in the Department of French and Italian at the University of Iowa. He has published an edition of the *Descriptions poëtiques* by Jean de Bussières (PFSCL) and an edition of the *Violier des histoires rommaines* (Droz). Currently, he is working on an anthology of poetry for students who read French at an advanced level but who are still novice readers of poetry.

David LaGuardia is an Associate Professor of French and Comparative Literature at Dartmouth College. He has published on the French *nouvelle* (*The Iconography of Power: The French Nouvelle at the End of the Middle Ages*, University of Delaware Press, 1999), Rabelais, Blaise de Monluc, and Marguerite de Navarre. He is currently at work on a manuscript devoted to masculinity in sixteenth-century French literature, and is planning a long-term project on personal memoirs of the period.

Jeff Persels is currently Associate Professor of French at the University of South Carolina. His publications include articles on gender issues in Rabelais and Marguerite de Navarre and on the polemical uses of bodily functions in vernacular French Reform writings.

Josef Schmidt is Professor of German Studies at McGill University. Recent books include: *On the Genealogy of Antisemitic Nazi Jargon in the German Speaking Countries. The Catholic Contribution* (in progress); *Quo vadis? – woher kommst du? Unterhaltungsliterarische konfessionelle Apologetik im Viktorianischen und Wilhelminischen Zeitalter*, Canadian Studies in German Language and Literature, vol. 39 (Bern: Lang, 1991); with Marion Moamai, *Marie de l'Incarnation*: *'Der Lebensbericht von 1654'*, in *Auswahl übersetzt und eingeleitet. Christliche Meister* 28 (Einsiedeln: Johannes Verlag, 1986).

Mary Simon is a Bibliographer/Collections Development at the Osler Library of the History of Medicine, McGill University. Her publications include *Disease in History: Medical Profiles of Illustrious Persons*. (Montreal, QC: Osler Library of the History of Medicine, McGill University, 1996).

Peter J. Smith is Senior Lecturer in English at Nottingham Trent Unviersity. He is author of *Social Shakespeare* (Macmillan 1995) and the co-editor of *Hamlet: Theory in Practice* (Open University Press 1996). He has edited Christopher Marlowe's *The Jew of Malta* and *Edward the Second* as well as Thomas Dekker's *The Shoemaker's Holiday*. He is Associate Editor of the international journal of Renaissance Studies, *Cahiers Elisabethains*.

Alison G. Stewart is Associate Professor of Art History in the Department of Art and Art History at the University of Nebraska–Lincoln. Her publications have dealt with Northern Renaisssance secular art and peasant festival imagery from *Unequal Lovers* (1979) to articles on fountain of youth and nose dance prints and two forthcoming books: *Before Bruegel: Sebald Beham and the Origins of Peasant Festival Imagery* (2005) and *Saints, Sinners, and Sisters* (2003), an edited volume addressing gender and art.

Joseph Tate is Adjunct Instructor at Lewis & Clark College. His recent publications include, 'Numme Feete: Meter in Early Modern England,' *Listening to the Early Modern*, a special issue of *EMLS* (forthcoming); and 'The Motion of Emotion: Physical Responses of Early Modern Playgoers,' *Shakespeare in the Classroom* (Fall 1998).

Emily E. Thompson is Associate Professor of French at Webster University (Saint Louis). She has published articles on Bonaventure Des Périers and is working on a book focusing on alternate discourses in the French *nouvelle* tradition.

Jeanne Morgan Zarucchi is Professor of French and Art History at the University of Missouri–St. Louis. She has published articles on medals in *Studi Francesi*, *EMF*, *Seventeenth-Century French Studies*, and *Source: Notes in the History of Art*. Zarucchi has authored numerous studies of the life and works of Charles Perrault, including a critical edition of his *Memoirs* (University of Missouri Press). She has also researched works by the seventeenth-century artists Artemisia Gentileschi (*Woman's Art Journal*) and Philippe de Champaigne (*Seventeenth-Century French Studies*).

Introduction

Scatology, the Last Taboo

O what lovely fecal matter!
François Rabelais

Where there is dirt there is system.
Mary Douglas

This collection of essays was provoked by what its editors considered to be a curious lacuna: the relative academic neglect of the copious and ubiquitous scatological rhetoric of Early Modern Europe, here broadly defined as the representation of the process and product of elimination of the body's waste products (feces, urine, flatus, phlegm, vomitus). Our most educated forebears, different from ourselves, did not disdain it – if such proof may be found in the mere proliferation of examples – and, further, employed it in all manner of works, not just in the crude jokes of comic ephemera. This neglect led to the idea of an anthology that would invite reconsideration of the many forms and functions of scatology as literary and artistic trope. The results emphasize that while the Rabelaisian corpus may yet serve as the standard referent, hallmark or even touchstone of the scatological in Early Modern European works, critical inquiry must move beyond this so that readers may extend and deepen their understanding of what the *Oxford English Dictionary* dismisses simply as 'dirty literature.'

Worthy children of a Classical, Romantic and, most tellingly, bourgeois aesthetic, we can hardly be blamed for several centuries of discomfort, in both our teaching and our writing, when faced with works that deal with that last taboo, what Victor Hugo evocatively called the 'last veil' clouding our vision of the truth.[1] Sexuality in all its myriad forms has long been the darling of academic readers, a once marginalized, now legitimate field of critical investigation, commentary and theory building. Scatology, however, arguably an even more universal function than sexuality, still retains the power to make us blush, to provoke shame and embarrassment. Discussion of excrement is generally relegated to one of two extremes: the objective, clinical discourse of medical and social sciences (e.g., gastroenterology, psychology, anthropology) or the subjective, gross indecency of infantile insult or juvenile jest (e.g., *South Park*). The contributors to this volume reconsider this last taboo in the context of Early Modern European artistic and literate expression, addressing unflinchingly both the objective reality of the scatological as part and parcel of material culture – inescapably a much larger part, a much heavier parcel then than now – and the subjective experience of that reality among contemporaries.[2]

If students of literature and the arts have hitherto and in the main been reluctant to tackle, or squeamish about addressing, scatology in earnest, a slowly growing number of recent works (e.g., Vigarello, Monestier, Inglis) have articulated for them and modeled, to varying degrees, socio-historical interpretations of excrement as process, product and experience. Such interpretations owe much to at least three distinct but arguably mutually compatible intellectual trends.

First, the ethnographic fieldwork and analysis of such anthropologists as Claude Lévi-Strauss (*L'homme nu*, 1971) and Bernadette Bucher (*Icon and Conquest*, 1981) but especially Mary Douglas (*Purity and Danger*, 1966) posit a symbolic connection between 'dirt' and 'danger' as the formative relationship of a given society's cosmology, the desired elimination of both in the search for 'purity' constituting then 'a positive re-ordering of our environment' (Douglas 2). For Bucher, as for Douglas, 'impurity,' and 'disorder' are synonymous. From a social standpoint, Bucher claims that 'what is decreed impure, [and] thus execrated and condemned by a culture, is an object out of place, a cause for disorder' (142). Excrement becomes part of this disorder and marginalization because it is both naturally present but, in most cases, socially absent. It finds itself in 'ambiguous and confusing' circumstances because it is of the body but then physically dislodged from it. Consequently, human waste is separated from the individual who created it, and from the society that rejects it. Paying close attention to this 'disorder,' understanding the treatment of impurity and its concomitant 'danger' within a given society's conceptualization of its own nature, becomes critical to a full and accurate appreciation of that society.

Second, the popular versus official cultural dichotomy of Russian theorist Mikhail Bakhtin attempts to do just that, focusing on the subversive, 'carnivalesque' nature of the grotesque body and its excrement-producing 'lower stratum' in the works of Rabelais (*Rabelais and His World*, 1965). His approach has indelibly marked scholarly readings of literary and artistic scatology, particularly that of Early Modern Europe, and its influence, as we might expect, is clear in the number of following essays that take it as a frame of reference. His early attention to the socio-historically specific culture of the 'main events in the life of the grotesque body, the acts of the bodily drama' (317), and the copious critical literature it has spawned would, in fact, seem to have all but eclipsed earlier perspectives. If Freud and psychoanalytical approaches to scatology were once the obvious interpretive choice for modern readings of primary texts such as those treated here – the standard set by Erikson's influential biography of one the Early Modern era's best-known scatologs, (*Young Man Luther*, 1958) – such is no longer the case. For a variety of reasons, many of them connected to Bakhtinian and New-Historicist attention to the recovery and explication of European 'popular' culture, the postulation of a psychological, ahistorical reading of human functions and the way individuals and groups in and across time and space perceive and interpret them has been necessarily modified. The contributors to this volume are all aware of and seek to understand the mental and physical distance that separates us from the experience of Early Modern excrement. What emerges from their work we may usefully define as a set of complementary applications – the first by the primary authors, the second by their modern readers – of Michel Foucault's idea of 'a transformation into discourse' as outlined in his *History of Sexuality*. His consideration of the 'censorship,' 'denial,' and 'repressive hypotheses' (12) used to thwart the development of human sexual identity finds a parallel with the scatological in that the social desire to silence literary and artistic representations of it translates into an aesthetic and linguistic code whereby the purgative becomes expressive. As Douglas similarly argues, 'The danger risked by boundary transgression is power' (161). Evoking reactions of disgust and/or ribald delight, the texts and illustrations under examination unleash creative forces and responses that alter our perception of what the form and function of art actually are. Cultural suppression becomes subcultural revelation as what was once rejected as waste is now valued as inspiration. Or, rather, as at least one critic has likewise argued in a corrective to Bakhtin, the distinction between high and low culture, like the rejection and subsequent recuperation of waste, actually corresponds more to the way we have chosen to recover the past than to any real separation acknowledged among Rabelais's contemporaries. As is the case in many of the Amerindians studied in Lévi-Strauss's *L'Homme nu*, their excrement was always already useful, recyclable, both literally and figuratively; part of the effort of the following essays is to make that point.

How that always already useful and recyclable Early Modern excrement was lost, so to speak, is the concern of the third trend. German sociologist Norbert Elias (*The Civilizing Process*, 1939) developed, with an acknowledged debt to Freudian psychoanalytical theory, the seminal notion of a historically documentable European 'civilizing process,' a process very much concerned with the scatological. Most pertinent to this collection, Elias zeroes in on what he considers to be the beginnings of an historical shift in modes of social behavior in Early Modern Western European society concurrent with the literary and artistic works examined in this volume. He founds his notion of a civilizing process on a gradual modification in 'personality make-up' or 'habitus' – including, but not limited to, those involving attitudes toward the excretory experience. Motivated by the rise of a 'courtly' and/or 'bourgeois' habitus, both of which became increasingly scandalized over time by that experience and, as a result, increasingly censorious of its representation, the shift can be readily documented in the rise and proliferation of manuals of conduct. It is worth noting here that all three trends focus on varying forms of private and public control of excrement and excretion – the overall 'excretory experience,' as one author would have it – as essential to a given society's cosmology, whether literal and physical or symbolic and moral. Elias's postulation of a 'civilizing process' for Early Modern Europe hinges on this.

Building on Elias, Douglas and the work of sociologist Pierre Bourdieu, David Inglis has most recently sought to incorporate an elemental 'ethnography' of dirt into the Eliasian scheme of civilization. Elias himself points the way, as we have seen, in linking the rise of an eventual 'bourgeois habitus' to self-conscious modifications in the codes of social interaction, of which those applying to the most 'unclean' and hence 'dangerous' of them all, excretory practice, are the most problematic and so subject to most rigorous control and even repression. This, as

Inglis traces most convincingly for the modern period – his 'history' from Antiquity to the late seventeenth century is disappointingly thin – has been effected in the West (his examples are primarily French and English) in both the public and private spheres. Dirt hence disorder hence danger become associated with the proletariat, the proverbial 'unwashed' even unwiped masses, as distinct from the hygienically sound hence orderly hence safe bourgeoisie – that is, until the former, too, come to adopt Inglis' 'bourgeois' now almost universal Western 'fecal habitus,' ultimately depriving dirt of its utility as a class distinction. The 'civilizing process' here becomes synonymous with the rigorous public and private effort to distance oneself from one's own excrement, the sight and smell of which grow proportionally offensive. That offense transfers easily to those words and images that represent that sight and smell, resulting in as much discomfort with scatology as with the excretory experience itself. Rabelais's 'bathroom humor' becomes the cause of an embarrassed snicker, the object of academic dismissal, the reason we read him in private but gloss over the 'dirty bits' in public. All the more so as he, like many of his contemporaries treated in this anthology, has the vexing habit of mixing an *altior sensus* with the quest for a perfect asswipe.[3] Much Early Modern vernacular art and literature is disorderly, is unclean, is thus 'dangerous,' subversive, and is in need of the neo-Classical bath it will receive in subsequent centuries.

Even more illuminating for the argument that links the essays in this volume, Elias's primary cultural marker, Erasmus – whose 1530 conduct manual, the *De civilitate morum puerilium* (*On Good Manners for Boys*), is an important milestone in the 'civilizing process' – not only announced the advent of the specific socio-historical scatological moment, as it were, that would become ours, but also that he was himself aware of participating in one. A curiously revealing case in point, the *Adagia*, compiled over the course of his career, explicate many a proverbial scatological act (of micturition, of excretion) toward which the commentator demonstrates a predictable – following the Eliasian thesis – and telling reserve. Yet adage 3.7.1, *Scarabaeus aquilam quaerit* ('A dung-beetle hunting an eagle'), acknowledges, as much as any other contemporary work treated in the following pages, a relationship to excrement different from our own:

> The fact that it [the dung-beetle] uses the droppings of animals for its own purposes is a matter of praise, not accusation. As if doctors do not do exactly the same, not only making ointments with a variety of animal and even human excrement, but prescribing it in medicines for the sick (297)[4]

Moreover, he continues, in explicit recognition of his own historically determined, and thus intrinsically mutable, relationship to the scatological:

> But is it also true that men are offended not so much by excrement itself as by the current view of it; to the earliest mortals this substance was not so disgusting as it is to us, for they called it by the very auspicious name of *laetamen* ['manure,' from *laetare*, 'to gladden'] and they had not hesitation in giving the god Saturn the nickname of 'Sterculeus' [from *stercus*, 'dung, shit'], and this was a compliment if we believe Macrobius. (298)[5]

Erasmus, both harbinger and codifier of a 'civilizing process,' of a new 'scatologically-challenged' habitus-in-the-making that would forever distance us from our excrement, noted himself, and with all the troubled 'objectivity' of an ethnographer, his own and his contemporaries' distance from an earlier scatological golden age, 'if we believe Macrobius.'

That this Erasmian/Eliasian shift coincides with the same historical moments and spaces inhabited by the works discussed in this anthology – works constitute so many witnesses to and agents of that change – is worth exploring as a hypothesis for dispelling some of the inevitable and discomfiting 'ambiguity' surrounding excrement, for clearing away, as it were, the taboo on serious treatment of scatology in art and literature. What 'clouds' our ability to appreciate the frequent Early Modern recourse to excremental rhetoric, whether in text or in image, is, as Erasmus suspected, our own socially, culturally and historically determined distance from an earlier scatological golden age. Traced anthropologically, sociologically, culturally and historically, the Early Moderns arguably shat differently (not to mention ate, drank, digested, pissed, farted, vomited and spat differently) as well as inherited and cultivated a different understanding of those paradoxically both natural and grotesque acts. Explorations, however tentative, of that difference should render Early Modern Europeans' less abashed use of scatology less ambiguous, less unsettling, more meaningful. Although far from comprehensive, the following essays on some of the period's cultural artifacts begin to do just that, looking for, to paraphrase Douglas, the system in the dirt, for '...if uncleanness is matter out of place, we must approach it through order' (40).

For ease of consultation, the editors have decided to group the essays geographically and chronologically with regard to authors and works treated. That a good half of them focus primarily on the French tradition – in inverse proportion, one might argue, to the importance of scatology as recognized in national stereotypes – is as much a reflection of the editors' own fields of inquiry as it is indicative of the relative lack of attention Early Modern French scatology has hitherto received, as opposed to the German and, to a lesser extent, English varieties.

Both Barbara C. Bowen and Geoffrey R. Hope, tracing the fortunes, respectively, of comic flatulence and of one specific political and moral excretory anecdote, cross national, generic and linguistic frontiers in pointing out important Classical and neo-Latin antecedents, influences and parallels in a subject most often associated with early vernacular 'earthiness.' The result opens up a large body of scatological material that was as familiar to the Humanist contemporaries of Rabelais as it is most likely unknown to, or certainly underappreciated by, modern readers. In both authors, the seeming off-color or embarrassing joke takes on unexpected rhetorical and social importance.

A trio of essays focuses specifically on sixteenth-century France and Francophone Switzerland. David LaGuardia takes on the former's most celebrated scatolog, François Rabelais, in a syncretic reading of *Pantagruel* and *Gargantua* that contributes to two major trends in Rabelaisian scholarship, the medical and anatomical subtext articulated by Rabelais the professionally-trained and practicing physician, first systematically approached by Roland Antonioli; and the eschatological 'design' drawn by Rabelais the Evangelical Humanist, most recently and methodically explicated by Edwin Duval. For both approaches, as LaGuardia demonstrates, the giant princes' digestion and its discontents, manipulated by contemporary dietary prescription and proscription, are critical to a full understanding of the chronicles.

Jeff Persels uses one particularly vivid anti-clerical anecdote of vomiting, exploited by Calvinist theologian Pierre Viret, as a springboard for discussion of the polemical uses of scatology. As a useful figure for referencing the critical controversy surrounding the Catholic mass, it is but one example of the evocative power of 'vulgar' language in framing many issues of contemporary doctrinal difference.

Emily E. Thompson shifts scrutiny from the religious and the medical issues of the preceding two essays to the social, glossing both the moral implications and the class consciousness of scatological anecdote in tales by Marguerite de Navarre and Philippe de Vigneulles. She argues effectively that figurative Early Modern use of the excremental was not limited merely to obscene humor nor to Evangelical proselytism. From a psychological standpoint, the scatological transforms pride into humiliation, thus emphasizing the personal and collective 'instability' Marguerite and Vigneulles saw as typical of Renaissance France.

Continuing into the seventeenth century, Russell Ganim reads against the critical grain in privileging the scatological motifs of the baroque as exemplified by Théophile de Viau's 'cabaret' poetry, verse that contemporaries (and not a few modern critics) of the *Grand Siècle* would have preferred to relegate to the preceding era of the less 'civilized' authors examined by LaGuardia, Persels, and Thompson. His consideration of Théophile's long-neglected 'crass' libertine works argues for their inclusion as a necessary component of the Baroque aesthetic, fleshing out the anatomy of the baroque body much the way Peter J. Smith does in the reading of English cavalier poetry that concludes this volume.

Jeanne Morgan Zarucchi uncovers and interprets an unknown 'scandalous' piece of political commentary on the reign of Louis XIV long hidden in a modified catalog of laudatory medals. A representation of the king shitting into a chamber pot held by the pope protests the 1689 Revocation of the Edict of Nantes, perpetuating the polemical utility of scatology, and to much the same ends, echoes the sixteenth-century Calvinist polemicists discussed by Persels.

Crossing the border into Early Modern Germany, both Glenn Ehrstine and Josef Schmidt survey and remap the familiar territory of Lutheran scatological rhetoric. Schmidt looks to common traditions in medieval piety and medicine – specifically the understudied *Dreckapotheke*, too innocuously translated as 'Filth Pharmacy,' he argues – for sources of popular expressions of Reform issues, many of which left a mark on the language down to the present day. Ehrstine takes on scatological motifs in the corpus of German *Narrenliteratur*, recalling and assessing physical cures for folly, many involving purges such as those LaGuardia examines in Rabelais. In this discussion of corporeal purification, Ehrstine comes to similar conclusions about the meaning and usefulness of bodily catharsis and its relation to rhetorical strategy.

Alison G. Stewart carries these concerns over into the visual arts, reading representations of vomiting, urination and defecation in peasant festival images from Germany and Flanders. Working from pre-Bruegel prints and paintings, she traces the waxing and waning of moralizing scatological motifs, seeking cultural explanations for their explicitness and ubiquity in German art and their understatement and scarcity in Flemish art.

Across the channel, Joseph Tate reviews the state-of-the-art Early Modern science of uroscopy, the specialty of the 'Pisse-Prophets,' as background for understanding the detailed diagnosis, prognosis and death of Christopher Marlowe's Tamburlaine. He thus grounds a new interpretation of Marlowe's original addition to the possible source materials for the play on a reading of contemporary medical and popular culture, restoring the importance of an overlooked but vital Early Modern scatological frame of reference. Finally, Peter J. Smith closes the volume with a speculative reading of English Cavalier verse that posits a shift in sensibility with regard to the use of scatological rhetoric effected between Sir Thomas Urquhart's 'jubilant' and carnivalesque 1653 translation of Rabelais and the dark, sterile scatological verse of the post-Restoration Cavalier poet John Wilmot, Earl of Rochester. Smith effectively alludes to the topos evoked by all essays in the volume, that of the scatological as a fertile trope of both renewal and decay.

It is perhaps not merely serendipitous that the collection of these essays should follow so closely on the English-language publication of Dominique-Gilbert Laporte's brief but meditative 1978 essay, *A History of Shit*, and Ralph A. Lewin's light yet informative *Merde: Excursions in Scientific, Cultural and Socio-Historical Coprology*. The latter's adjective-laden subtitle would, in fact, with the addition of 'literary' or 'rhetorical,' serve this volume equally well. These are essays informed as much by close attention to Early Modern 'sciences' – diet, hygiene, uroscopy, nascent etiology and pathology, even the German tradition of *Dreckapotheke* or 'Filth Pharmacy' – as by the associated psychological implications for language, poetry, narrative and the arts. There is decidedly both an individual and collective effort on the part of the contributors to account for the rhetorical recourse to images of physical elimination in contemporary terms, thereby enriching our understanding of many familiar works and seeking a place in the canon for some hitherto neglected or underestimated ones.

Historically we humans have gone to great lengths to render civilization synonymous with the marginalization of human waste and its production, restricting it to discrete corners of our lives and minds, banishing it from our educated, polite discourse. We might recall here that the publication of Erasmus' book of manners mentioned earlier, the 1530 *On Good Manners for Boys*, rubbed shoulders with Rabelais's 1532 *Pantagruel*: scatological expression was thus already marked for the suppression we 'cultivated' moderns seem to prefer, though as these essays show, it was a long and glorious decline. That decline may seem well and good to us now but it has not come without cost, part of which is arguably our willful deafness and blindness to the richness of the scatological as metaphor, to how the expression to 'cheat' came to be expressed as *bescheissen* or 'to shit upon' in German, to why knowing the mettle of a man by his urine could be expressed in French as *Je voudrois bien veoir de son urine*, that is, 'I would like to see his urine.' Achieving a new respect for, contributing knowledge to and fostering interest in Early Modern scatology within the realm of literary and art history studies would mean, without blush or shame, that this collection has been 'well shat' (*bien chié*). To each generation its idiom; for discerning readers and spectators to gauge its value without prejudice.

Notes

[1] Cited by Peter Stallybrass and Allon White in a chapter devoted to the semantics of the nineteenth-century sewer, 'The City: the Sewer, the Gaze and the Contaminating Touch,' *The Politics and Poetics of Transgression*: 125-48: 'But in describing the functional process of cleaning, [author of *London Labour and the London Poor*, 1861, Henry] Mayhew articulates the sewers as symbolic system. Indeed, he repeats one of the dominant tropes of western metaphysics: truth lies hidden behind a veil. But 'truth' is now conceived materially, as excrement. In *Les misérables*, in what might be called, without irony, one of the most brilliant explorations of the semantics of the sewer, Victor Hugo wrote that there could be no 'false appearance' in the 'vast confusion: of the 'ditch of truth': '[the] last veil is stripped away…' (140).

[2] Although far from numerous, serious historical considerations of the Western excremental experience, including accounts of the Early Modern period, do exist, and have been on the increase most recently, from an important nineteenth-century account of Paris by Alfred Franklin's 1873 'Étude sur la voirie et l'hygiène publique à Paris depuis le XIIe siècle' to David Inglis's 2001 *A Sociological History of the Excretory Experience: Defecatory Manners and Toiletry Technology*, and including George Vigarello's 1985 *Le propre et le sale: l'hygiène du corps depuis le Moyen Age* (translated as *Concepts of Cleanliness: Changing Attitudes in France Since the Middle Ages* in 1988), as well as Dominique-Gilbert Laporte's 1978 tantalizingly brief essay *Histoire de la merde* (tellingly translated into English only in 2000), and Martin Monestier's richly illustrated 1997 *Histoire et bizarreries sociales des excréments, des origines à nos jours*, which is much indebted to Franklin's pioneering work.

[3] Cf. especially folklorist Claude Gaignebet's *A plus hault sens: l'ésotérisme spirituel et charnel de Rabelais*.

[4] *Adagia* 3.7.1.486–89: *Porro quod animantium excrementis ad suam commoditatem abutitur, ingenii laus est, non crimen. Quasi vero non idem faciant medici, qui cum multorum animantium tum hominis etiam excrementa non illinunt solum, verum in potione ministrant in morbis.*

[5] *Adagia* 3.7.1.499–503: *Quonquam homines quoque non tam res offendit quam opinio; nam priscis illis mortalibus res ipsa non perinde atque nobis visa est abominanda, quam auspicatissimo vocabulo laetamen appellarunt. Nec dubitarunt Saturno deo Sterculei cognomen addere, nimirum honoris causa, siquidem Macrobio credimus.* The Macrobius reference is to *Saturnalia* 1.7.25: *Hunc Romani etiam Sterculium vocant, quod primus stercore fecunditatem agris comparaverit* ('Moreover, at Rome men call him 'Sterculius,' as having been the first to fertilize the fields with dung' [59]).

Chapter 1

The 'Honorable Art of Farting' in Continental Renaissance Literature

Barbara C. Bowen

Literature, from Antiquity to some time in the eighteenth century, was frequently characterized by a simple juxtaposition of two Latin adjectives: *utile* ('useful or profitable') and *dulce* ('sweet or pleasant').[1] In other words, literature aimed both to please and to instruct, like the rhetoric which underlay it, and comic literature could be just as 'useful' as tragic or openly didactic literature. So even scatological comic literature (which has been very little studied because scatology was until recently a taboo subject) could have satirical, rhetorical or sociological implications.

This is a vast topic, and the main aim of this article is to draw attention to it, rather than to attempt to cover it. I have chosen to consider scatological humor, and specifically farting, in examples from three related Renaissance literary domains: anthologies of *facetiae* in Latin, Italian and German; French comic narrative; and parodies of scholastic argument in Latin and French.

Neo-Latin *Facetiae* Collections

Despite the efforts of a few critics, myself included, Renaissance anthologies of *facetiae* (which can very roughly be translated 'jokes') are still not well known or widely read.[2] I am talking about at least 32 collections (my list is certainly not complete) written, and for the most part published, between the 1340s and 1600, in Latin, Italian, French, English and several varieties of German. In many cases, their title contains the word *facetiae* or a vernacular equivalent (*facezie*, *motti*, *facecies*, *mery talys*), but in many cases it does not; one would not guess that works called *Mensa philosophica* (*The Philosophical Table*) or *De cardinalatu* (*Concerning the Cardinalate*) include joke collections, nor, probably, that there are jokes in Leonardo da Vinci's *Notebooks* and in Erasmus's *Colloquies*.

A few of these works were compiled by well-known names: Petrarch, Poggio, Leonardo, Bebel, Erasmus, Castiglione, Thomas Wilson. And they all owe something to the Ciceronian rhetorical tradition, whether or not they refer explicitly to Cicero's distinctions in the *De oratore* (*On the Orator*) between *facetiae* ('witty remarks') and *sententiae* ('serious remarks'), jokes *in re* ('stories') and *in dicto* ('plays on words'), and so on. This Ciceronian tradition, kept alive by Macrobius in the fifth century, apparently defunct in the Middle Ages but revived by Petrarch in the fourteenth century, was astonishingly tenacious; in the late sixteenth century a French medical authority, Laurent Joubert, is still using Cicero's terminology to categorize humor in his treatise on laughter. So we would not really expect Renaissance *facetiae* to be scatological at all, since Cicero was quite explicit that *obscenitas* is never permissible, either in official oratory or even in dinner-table conversation.

Petrarch undoubtedly agreed with Cicero, but fifteenth- and sixteenth-century Humanist writers did not, judging by their frequent reference to buttocks, farting, laxatives, defecation and excrement. It seems that taboos were similar to ours today: in stage farces characters quite often pee, but very seldom defecate, on stage; farting in public, apparently more common than today, as we shall shortly see, could be deliberately induced (by men) or was something to be ashamed of (for women); and we know that readers were shocked by Alciati's emblem of a young man defecating into a cooking pot, which was an explicit analogy of homosexuality. On the other hand Erasmus, who disapproved of obscenity quite as much as Cicero did, retells a joke claiming that the bottom is the most honorable part of the human body – because it gets to sit down first.[3]

We must beware of generalizations about the joke-books, but it seems fair to say that, with a few exceptions, 'dirty' jokes are *either* sexual *or* scatological, that the German collections in general emphasize scatology more than sex, and that the French compilers were more prudish than either the Italians or the Germans. Most Latin, Italian and German anthologies include a number of scatological jokes, usually more about defecation than about farting. Since Latin was the language of the international scholarly community, the Latin collections were no doubt the most influential.

Poggio, author of the first important Latin anthology (*Facetiae*, 1470), unusually includes exactly the same number of farting jokes (six) as of jokes about excrement.[4] Farting can be perceived as comic insult, as when the Cardinale di Tricarico insults Alto de' Conti by farting at him (this is a *facetia iucundissima* 'very delightful joke') (136: 262), or the Cardinale de' Conti, at a picnic on a hot day, asks a secretary to make a breeze for him; he farts, and all those present laugh (135: 262). Or it can be a pretext for sarcasm: the (real) fool Gonnella, asked by a man from Ferrara to make him a diviner (*divinum*) in exchange for money, shares a bed with him, farts, and asks him to put his head under the bedclothes. The man says, 'You farted,' and Gonnella replies, 'Pay me; you're a diviner' (165: 294). This anecdote is titled *facetissimum* ('very funny story').

Poggio's last farting joke is one of the exceptions which combine sex and scatology (213: 344). A young wife sees sheep and rams copulating, and asks her husband how the ram chooses his sheep. The husband unwisely replies that he chooses the one that farts; the wife then asks if this also holds good for people. When the husband says yes, she immediately farts, with the expected result, and shortly thereafter does it again. But when she farts for the third time, the exhausted husband says: 'I'm not making love to you again, even if you shit out your soul' (or 'heart': *Non si cor cacares*).

The next most 'popular' Latin collection was Heinrich Bebel's *Facetiae* (1508–12). The two anthologies have a good deal in common; both authors were well-known Humanists attempting to put into good Latin earthy anecdotes originally told in Italian or German. Bebel includes a total of 18 stories about urine, excrement or vomiting, including one about a priest who shits in the baptistery (3.33: 117), and one in which a court fool shits in a prince's greaves (3.145: 158), but only three about farting. In 2.60 a citizen of Rottweil is about to address the Emperor Sigismund when he suddenly farts. Instead of being disconcerted, he looks round at his bottom and says to it: 'If you want to speak, there's no need of my oration,' and calmly continues his speech (66). The Emperor, who has a good sense of humor (*qui hilaritate gaudebat*), is delighted.

In 2.74 a civic magistrate is addressing Countess Mathilda of Austria, on some official occasion, when he suddenly farts. Not at all disconcerted, he continues speaking, and the countess pretends she has heard nothing, but the attendant girls standing behind her all laugh, so much so that one of them farts. At this, the magistrate turns round and says: 'Proceed in order! And when it's my turn again, I'll start a new round.' Everyone bursts out laughing, and the countess is so pleased that she treats the magistrate very well (*magnifice eum tractavit*) (70–71).

Joke 2.123 is an intriguing story, one of many of Bebel's which re-create the atmosphere of contemporary German church services. The baptism service apparently begins with an exorcism, and a priest baptizing a baby boy is reading the biblical passage where Jesus spits on the ground and makes clay of the spittle (John 9.6). At this point the woman holding the baby, bending to pick up some dust from the ground, emits a horrendous fart. The priest seizes the opportunity to announce that the sound was the Devil going out of her, thanks to the priest's words. The woman, however, overcome by shame and not having heard what the priest says, claims that it was the baby, not she, who farted. Upon which the priest, annoyed, asks God to curse the baby, since given his impudence at such a young age, what will he do when he is old (89)?

Johann Gast's *Convivalium sermonum liber* (*Book of Sociable Conversations*) of 1541 is the first of the big sixteenth-century anthologies; the first edition contains over 400 *facetiae*, of which 11 concern urine or excrement and only three, successively numbered, concern farting.[5] No. 93 tells how the Emperor Claudius considered issuing an edict permitting farting at banquets (*in convivio*), on the grounds that holding back a fart for reasons of modesty can be dangerous (a reminiscence of Suetonius's *Life of Claudius*), and no. 94 repeats Bebel's 2.60 about the man addressing Emperor Sigismund.

Gast's no. 95 is a lengthy story about a rich man in agony with colic and begging God and all the gods (!) first for relief – 'a fart' (*uentulum uentri*) – and then for death. At which his fool (*morio*) calls him a fool (*fatuum*), for asking God and the gods for the Heavenly Kingdom when they will not even grant him a fart. The rich man bursts out laughing, and farts; the doctor says that fart is worth a thousand gold coins; the fool then emits a much louder fart and inquires how much that one is worth.

Nicodemus Frischlin's *Facetiae selectiores* (*Well-Chosen Jokes)* include two fart jokes. One (13) is another witty retort by an involuntary farter: 'My bottom wanted to say something, but lacked your tongue'; in the other (38), an ignorant novice thinks *pedum*, which is the Latin word for a bishop's staff, means 'fart' (Latin *peditum*).[6]

Vernacular *Facetiae* Anthologies

The vernacular anthologies tend to be less outspoken on taboo subjects than the Latin ones. The first Italian joke collection, and in fact the first vernacular joke-book, is Lodovico Carbone's *Cento trenta novelle o facetie* (*130 stories or jokes*), 1469–71(?), not published until 1900. He includes no references to excrement, only two to urine (88 and 89), and one to farting (63), which is Emperor Claudius's proclamation, later referred to by Gast.

Probably the most curious of all these collections is that attributed to the priest Arlotto Mainardi, *Motti e facezie del Piovano Arlotto* (*Witticisms and Jokes by Arlotto the Priest*), ca. 1480. Its 'jokes' run the gamut from rude anecdotes to moral fables presenting Arlotto as a Christ-like figure, and several include references to excrement. One (200) relates how the priest and his friends, while out walking, pause to pee, and Arlotto looses a terrible fart. The others marvel at this, but he says: 'Don't you think a trombone sounds pretty good among all these pipes ('pifferi')?' This is a good example of Arlotto's skill at putting down people who try to make fun of him.

The largest and most often reprinted joke-book of the sixteenth century is Lodovico Domenichi's, which went through numerous and varied editions from 1548 on. The 1574 edition, titled *Facetie, motti, et burle, di diuersi signori et persone private* (*Jokes, Witticisms and Gags, by various Nobles and Private Persons*), contains 981 jokes, about 30 of which are scatological, but only four are about farting. In this edition, most jokes are followed by a brief gloss in parentheses, whether by Domenichi himself or by an editor we do not know. Joke 1.60 is a close translation of Bebel's 2.60 about the witty speaker to Emperor Sigismund; curiously, the comment on this one is disapproving: 'Sfacciato, & da persona uile' ('Shameful, and by a base-born person') (26).

Also in book one, Domenichi retells Erasmus's joke about the *culo* ('bottom') being the most honored part of the body, but this time the anecdote has a conclusion: on a later occasion, when the learned men who had previously discussed which part of the body is 'la più honorata' are together again, the joker who had made the claim for the *culo* greets a colleague, who promptly farts at him on the grounds that he is saluting him with the body's most honorable part (102: 46). The comment on this one is 'Cauilloso'; we are not told why this one is witty when the previous one is insolent or shameless.

In the second book (2.98) we find a variation on Poggio's no. 213, though attributed to a contemporary. A woman asks Dardano Acciaiuoli what a dog mounting a bitch is doing, and he replies that it is preventing the bitch from farting. Naturally, she immediately tells him that she needs to fart, with the same results as in Poggio, and the comment prudishly deplores the sexual insatiability of women (101).

Domenichi’s last fart joke (7.13) is another witty exchange. Some students in the streets of Pavia see a working man peeing, and mock him by asking if he needs a knife (a comment on the size of his penis?). He asks if the knife they are offering cuts well, and when they say, ‘Yes,’ he ‘unsheathes’ (‘sguainò’) a fart, and tells them to peel it with their knife (401). This one, unusually, has no comment.

We might expect German collections to be easily distinguishable from Italian ones, but we are obviously dealing here with an international *sodalitas* of Humanists, similarly trained and with similar literary tastes. Jakob Frey’s *Gartengesellschaft* (*Garden Company*) of 1556 includes four farting jokes, two lifted from Bebel (70: baby’s nurse farts at baptism, Bebel 2.123; and 100: *burgermeister* farts in presence of Countess of Austria, Bebel 2.74), and one from Poggio (78: nobleman farts to give cardinal air on hot day, Poggio 135). Frey considerably expands this last story, explaining that the nobleman was German, not one of the cardinal’s servants, and therefore not obliged to do him a favor. The proper names are omitted, and the initial description of the cardinal more pejorative.

Frey’s joke 103, which I have not seen elsewhere, describes an old woman, in a company of young people of both sexes, who bends down to pick up things she has dropped, and emits a terrific fart. She excuses herself to the young women (‘it’s the fault of my age!’), but one of the young men says it is the fault of her witch’s asshole, and wishes that the Devil may go into all old witches’ assholes. At this she crosses herself at least eight times ‘für den arss’ (‘for the butt’) and wishes for God to go in there, and the insults accelerate on both sides (119). The import of this one is not clear, unless it is that the usually despised old woman holds her own against the insulting young men.

I am not familiar with all the sixteenth-century German joke-books, but there are two rather charming fart stories in Martin Montanus’s *Gartengesellschaft* of about 1560 (though none in his 1557 *Wegkürzer* [*Way-Shortener*]). In joke 43, a young bride in bed on her wedding night needs to fart, and is so terrified that her husband will smell it that she pulls up the bedclothes and tries to fart discreetly outside the bed. Her husband, who was pretending to be asleep, lets go with a much louder fart and says: ‘My love, let this one out of the bed too!’ (295). The poor bride is very ashamed (one wonders why, since it seems like a reassuring joke to make).

Joke 49 is a long story about the Devil’s attempt to win the soul of a farmer, who turns out to be smarter than he is. He demands that the Devil, in exchange for his soul, do three things for him, the third of which is to grab his fart and sew a button on it. Discomfited, the Devil thereafter leaves him alone (300).

This small, but I hope representative, sampling includes fart jokes which are just for fun, as well as those with some ‘useful’ result. Farts are always comic, at least to the bystanders, even in the case of the therapeutic farting recommended by the Emperor Claudius (Carbone and Gast), and illustrated by Gast’s rich man. Deliberate farting is either an insult (Poggio and Frey), or an instrument of revenge (Domenichi’s *facchino* puts down the students mocking him, and Montanus’s farmer routs the Devil). Involuntary farting, or talk about farting, can be a vehicle for satire of priests (Bebel and Frey), monks (Frischlin), cardinals (Poggio and Frey), women (Poggio and Domenichi) or doctors (Gast). It can also be a pretext for wit, either in a social context (Arlotto pays back the laughter at his fart with a snide comment on penis size; Domenichi's joker is put down by a fart), or in a rhetorical context. Bebel's man addressing Sigismund takes advantage of his fart to make a joke which endears him to the joke-loving emperor, while in a similar Bebel story the orator is addressing a woman, the Countess Mathilda, with equally happy results. In both anecdotes, picked up later by Domenichi and Frey respectively, an involuntary fart serves as an effective *captatio benevolentiae*; we shall meet a connection between scatology and rhetoric later in this chapter.

French Comic Narrative: Rabelais, Verville, Tabourot

We have, clearly, been discussing up to now both jokes and comic narrative, which can be differentiated only in cases where the work's title makes the distinction, as the *facetiae* anthologies sometimes do. Sixteenth-century French authors apparently preferred longer short stories to briefer jokes, and on the whole used less scatology than their Italian and German counterparts. I should like to consider here three French works, ostensibly comic narratives, which in their very different ways are related to the joke collections.

Given François Rabelais's reputation for gross language and anecdote, we are not surprised to discover over sixty references to farting in his five books of giant 'chronicles.' But whereas we might expect him to use stories from the Humanist joke collections, he never does so. He mentions the Emperor Claudius (*Œuvres* 580), and an unspecified pope who permitted everyone to fart (in Baisecul's incomprehensible *coq-à-l'âne* speech) (255), and the therapeutic value of farting is suggested by Panurge in his praise of green sauce (360). Apart from these passing references, there is no discernible connection with the *facetiae* anthologies we have been discussing.

Nor are there entire episodes based on farting, with the exception of the island of Ruach (638–42), whose inhabitants live on wind and die farting.[7] In chapter 27 of *Pantagruel*, Panurge jumps up and farts (310); when Pantagruel imitates him, his farts engender male and female dwarfs, whom he calls pygmies and settles in an island nearby – this is undoubtedly the most 'positive' farting I have come across.

One of Rabelais's specialties is the mixing and intertwining of literary and intellectual contexts. How better to spoof the epic than to have chivalric giant heroes farting, as in the example just given? But in fact the scatological is associated almost exclusively with the young Pantagruel and the child Gargantua; once both giants become Christian heroes there is no longer any mention of scatology, and farting most often signals either satire, or emphasis on the human and domestic by contrast with the epic or otherworldly.

Satire is usually obvious: Panurge sneers at the dilapidated walls of Paris by claiming that a cow could demolish them with a fart (268); Homenaz, the bishop of the despised Papimanes, laughs so hard that he hiccups, farts, drools and sweats at the same time (664); the satirical titles of the Library of Saint-Victor (236-41) include references to farting in both Latin and French (my favorite is *The Art of Farting Honorably in Society* [236]); and the evil spirit who has taken possession of Jacobe Rodogine makes her fart to indicate his refusal to answer questions (675). Panurge, who really does not want Pantagruel to cancel his debts, complains that people farting customarily say: 'That's for those who are out of debt ('les quittes'),' and so from now on all farts in the region will be aimed at him (369); and during his 'consultation' with the deaf-mute Nazdecabre (414), he makes a fart-like noise of derision ('la petarrade'). Rabelais twice uses the satirical expression 'you might as well get a fart from a dead donkey,' meaning 'what you are trying to do is pointless' (45, 466). In the latter case Panurge, once again, is venting his frustration with another unsatisfactory 'consultant,' Trouillogan the philosopher.

It is no accident that a majority of the farts, and references to farting, are attributed to Panurge, the very earthy, all-too-human foil to the sometimes superhuman Pantagruel. Not that emphasis on the human need be pejorative; Epistemon's return to the living after his death in battle proceeds by stages: first he breathes, then he opens his eyes, yawns, sneezes, and finally farts – at which point Panurge says: 'Now he's definitely cured' (322). But more often, farting does point up the grossly corporeal in a deliberately disgusting manner, as when the unfortunate old lady in Panurge's fable is being treated by the fox, who thinks her natural orifices are wounds; she farts so hard she stinks like a hundred Devils (271). Tappecoue's horse also farts in terror at the noise made by Villon's troupe of stage Devils (569), and when Thaumaste, the learned man from England, during his sign debate with Panurge, is getting the worst of the argument, he first farts involuntarily and then beshits himself with anxiety (288).

The explicitly down-to-earth tone of all this is more obvious in the French text, where a fart often has qualifiers: 'un gros pet' ('a great fart'), 'un gros pet de boulangier' ('a great baker's fart'), 'un gros pet de mesnage' ('a great household fart'), and where the verbs used of the terrified old lady are *vesner* and *vessir*, apparently considered even more vulgar than *péter* (the name of one of the litigants with whom Pantagruel has to deal in book two is Humevesne or Suckfart). As in the joke-books, so in Rabelais all farts are fun, but rather fewer of them are for profit; perhaps only Pantagruel's procreative farts could be so called! They are, however, undoubtedly part of a deliberately gross, earthy tone which helps to veil the serious, Humanist and Evangelical, message.

If there is one book in which we would expect to find more farting than in Rabelais, it is surely the enigmatic *Moyen de parvenir* (*How to Succeed*; probable author Béroalde de Verville, probable date of composition 1591, probable first edition 1612).[8] This pastiche of the literary symposium tradition, in which close to 400 speakers tell dirty stories in 111 bizarrely-titled 'chapters,' still baffles modern critics, but it is clear that obscenity and scatology, and more particularly the terms used to recount and discuss them, are the author's main preoccupation. (I once calculated that of the 355 pages of the Moreau-Tournon edition, only 36 have not one word likely to offend 'delicate' ears.)

There is much less scatology than sex (especially talk about sex) in the *Moyen de parvenir*, and I suspect less farting than excrement. In chapter 92 (302), 'Bail,' the game called *pet-en-gueule* ('fart-in-throat'), which was already on the child Gargantua's list of names (Rabelais 62), is mentioned by Saint Cosmas, who sees it as a symbol of a cyclical alchemical process.[9] Chapter 57, 'Sommaire,' is a long linguistic discussion in the course of which Alcibiades explains that in Babylonian *putain* ('whore') meant *madame*, and *vesse* ('fart') meant lady-in-waiting – or priest's concubine (173). Later he claims that priests' women can be called *vestales*, only *vessailles* is easier to pronounce, just as *vesses* ('farts') is easier to say than *vestes* ('jackets').

The rambling conversations of chapter 91 ('Doctrine') include a periphrasis for fart: 'a wind freshly laid' (like an egg), and some more reflections on terminology (294). Why, asks Brentius, if you call a man a *sot* ('fool'), does he assume you mean cuckold, and if you call a woman *vesse* ('fart'), she thinks you are saying whore? Poggio points out that if you call a man *pet* he will not mind, and that the difference between *pet* and *vesse* is that the first makes a noise (like a man importuning a woman) while the second is silent (like a woman refusing).

Nearly every 'chapter' includes a number of anecdotes, sometimes only a paragraph long, with more or less relevance to the subject ostensibly under discussion. In chapter 81, 'Instance,' Asclepiades recalls a servant girl carrying an egg, who suddenly needed to fart, but instead of tightening her buttocks to prevent it she clenched her hand, crushing the egg, and opened her buttocks, releasing a 'gros pet' (257). In the next chapter, 'Production,' we hear briefly about a Frenchman who farted in his sleep, and a Genevan who curses him, claiming that the French all fart like pigs (262), and in chapter 100, 'Attestation,' Thucydides tells of a 'good woman' who, bowing as she presented a loaf of bread, loosed a fart; all those present laughed, and she, thinking they were making fun of her small loaf, said: 'Pardon me, ladies and gentlemen, next time I'll make a bigger one' (319).

The only full-length story about farting is in chapter 7, 'Couplet.' A French gentleman is in bed with a Roman courtesan, who has prepared a number of little phials containing perfume. When she breaks one, her partner assumes it is a fart, and is astonished to smell the sweet scent; she tells him that Roman ladies' farts always smell like this. When she has repeated the maneuver a few more times, she accidentally does fart, and the gentleman of course dives into the bed to smell it; there follows a detailed description of how disgusting the odor is. She, not at all disconcerted, tells him that was just a little 'galantise' to remind him of his own country (14).

The *Moyen de parvenir* is frequently reminiscent of Rabelais, and occasionally of the joke collections, but it contains fewer references to farting than they do; farting is merely one of the scabrous topics discussed from time to time, and obviously less important than either excrement or sex. As we have seen, there is no therapeutic farting, and no discernible satiric intent; serving-women fart, but so do high-class prostitutes, and in any case the author is as interested in playing on words meaning 'fart' as in telling stories about them. So Béroalde de Verville, if he is indeed the author, definitely considers farting as something amusing (*dulce*) rather than profitable (*utile*).

My last French example is in several ways the most interesting of the three. Etienne Tabourot, author of the fairly well-known *Bigarrures*, published in 1588 the *Escraignes dijonnoises,* which have been much less studied because of their content, with which the prudish French have little sympathy.[10] It is a collection of 50 short stories, nearly all of which are either obscene or scatological, and 11 of which are about farting, including the first three, the last one, and the one adjudged the best (32). The title refers to the structure in which the storytelling takes place: a rustic shelter made of poles covered with sod and manure, where young countrymen and women gather in the evening. So most obviously we have here a pastiche of the traditional short-stories-with-frame, made popular by Boccaccio's *Decameron* and Marguerite de Navarre's *Heptaméron*, among many others. But there is more going on here, which the farting stories help to explain.

Several of these anecdotes recall the plots and characters of the joke collections, but I think only one is a direct imitation, the winning no. 32, which has exactly the same plot as Poggio's no. 213: wife, in her husband's absence, falls for a cowherd, who tells her a bull is mounting a cow because she farted; she propositions him, with the same results, until the fourth time, when even the cowherd's retort is the same as Poggio's: 'Don't fart any more, please, for even if you shit in the bed I won't do anything else today.' Two anecdotes use farting as a pretext for word-play. In no. 2, the speaker's aunt, Laurence, has a servant girl who farts, and claims that it was not she but the distaff (208); Laurence retorts: 'Does your distaff have a bottom ('cul')?' In no. 14, a husband-and-wife-in-bed story, the conclusion is : 'they made peace' – a play on *pets* and *paix* (211).

These stories call a fart a fart (*pet* or *vesse*), but most others use metaphors, which are either musical or military except in no. 50, where a fart is the West wind (226). In no. 1 a shy girl emits a 'petit sansonet' (208) which, according to Cotgrave means 'a little, or prettie and tunable, fart,' as well as a songbird, which is its meaning today. In no. 3, a 'belle jeune et délicate damoiselle' ('beautiful and delicate young lady') emits a fart so 'gentil et délicat' ('nice and delicate') that it sounds like a musical note (208); this is satirizing upper-class women with tight clothes.

Three stories liken farts to gunfire. In no. 20, a wife in bed emits 'deux ou trois cannonnades bien puantes' ('two or three really stinking volleys of cannonfire'), and her husband retaliates by peeing on her while quoting the adage 'a little rain calms a strong wind' (213; twice quoted by Rabelais [218, 255]). In no. 31 a husband's *canonnade* terrifies his wife during a thunderstorm (218), and in no. 34 a group of churchmen (!) play at extinguishing a candle ('canoner') with their farts; one of them farts so energetically that he shits all over the candle and the room (219). *Canoner* is particularly appropriate for church people subject to Canon Law.

While undoubtedly told for fun, these stories are often satirical, and in one case (no. 6) farting is definitely profitable. In a cobbler's shop, a man asks the price of a pair of boots, and is told, '50 sols,' at which he exclaims: '50 farts!' The cobbler says that if he can produce 50 consecutive farts, he may have the boots for free – and he does (209).

As these examples show, Tabourot is more interested in farting in domestic than in social contexts, and less interested than the joke collections in farting as a pretext for wit. This does not mean, however, that we have lost sight of rhetoric here; the whole work is a pastiche of, among other things, the Ciceronian rhetorical tradition. This is clear as early as the Prologue, which describes the *escraigne* in some detail, pointing out that its 'architecture' will not be found in Vitruvius or du Cerceau, and that it resembles a swallow's nest more than anything else. But a swallow's nest *is* a work of architecture, constructed according to certain formal principles. Are we setting up 'natural' rhetoric in opposition to 'artificial' rhetoric?

The terms used to discuss the stories include 'honeste' and 'gracieux,' implicitly comparing their scabrous content to that of the *Decameron* – or Castiglione's *Courtier* – and the village protagonists are intended as contrast to the 'ladies and gentleman' described by Boccaccio or Marguerite de Navarre. One of the basic principles of Ciceronian rhetoric is *copia*, abundance; what better joke than to compose an entire story sequence largely based on the abundant outpouring of excrement?

The *Escraignes* end with a facetious riddle session, which includes the following: 'What walks in the room, but has neither feet nor other members? A fart'; 'Who is the world's most inefficient crossbowman? The bottom, because it aims at the heels and strikes the nose'; 'Who emitted the first fart in Paris? The bottom.' And the author concludes that he has faithfully reported the *Escraignes* proceedings so that his gentle readers ('gens de bien') may be edified (!), or at least may take pleasure in their reading (226–7).

Parodies of Scholastic Argument

If Cicero would have shuddered at most of the material presented so far, he would hardly have survived reading the remaining works to be discussed here, which take Tabourot's use of scatology to spoof rhetoric to its logical conclusion. I can offer here only three examples of what seems to have been a discernible intellectual trend.

Two examples come from early seventeenth-century Paris. Bruscambille, whose real name was Des Lauriers, was an actor at the Hôtel de Bourgogne who specialized in satirical prologues, which owe a good deal to Italy, but also to Rabelais and the late medieval *sermon joyeux*. We have no means of knowing whether the published version faithfully reproduces the oral performance, but among his works titled *Paradoxes* we find: 'That a fart is something corporeal,' 'That a fart is spiritual,' and 'That a fart is a Good Thing.'

Contemporary with Bruscambille, and available in a nineteenth-century edition, is Tabarin, whose farces, dialogues, 'questions' and numerous other works served to attract the public to the Pont Neuf, where the charlatan Mondor sold his drugs.[11] Again, we cannot evaluate the authenticity of the published text, but we can say that these pieces are very often preoccupied with excrement. Among the *Questions*, which are short dialogues between Tabarin and his 'Master,' we find: 'Why do old men fart ('petent et vessent')?' (1: 50); 'Why do we fart while peeing?' (1: 83); 'Why do women cry and fart so often?' (1: 93); and (again) 'Why do we sometimes fart while peeing?' (1: 158). Similarly, among the *Fantaisies et Dialogues* we find: 'Why women fart more often than men' (2: 68), and 'What is the most joyful thing in the world?' (the answer is: 'a fart') (2: 94). Among the *Opuscules* is a list of recipes, one of which is to make people fart (grind up dried chestnut flowers and add them to the soup) (2: 292).

The dedicatory epistle to this pot-pourri, which was signed 'Tabarin' in the 1623 edition, begins with an astonishing sentence, 20 lines long, and containing by my count a total of 19 metaphors (1: 3–4). They include: protecting the book with the parasol of your (the addressee's) name; sheltering it under the roof of your intellect; passing the plane of your judgment over the bumps of its imperfections; barricading it within the fortifications of its incapacity against the cannon-fire of the envious, who will try to mix in the (ship's) deck of their conceptions in order to remove the poop and the masts of its doctrine, doctrine which has been pounded in the mortar of your intelligence.... And the author continues, apparently tirelessly, for another page. This is one of the best spoof dedications, and one of the best exercises in *copia*, that I have ever seen.

But by far the most elaborate parody of rhetoric and logic on this theme that I have found is a *Discursus Methodicus de Peditu, ejusque speciebus, crepitu et visio (Methodical Discourse on Farting, and its Species, Sound and Appearance*), one of several such parodies in prose and verse assembled in the *Facetiae Facetiarum, hoc est, Joco-seriorum fasciculus...* (*The Best of Jokes, or, A Booklet of Partly Serious Jokes...*) (73–98).[12] This work includes poems on the art of joking and the art of drinking, an elegy on drunkenness, a dissertation on kisses, and several pieces on women. I have not been able to discover who compiled this volume, or exactly when; there are several seventeenth-century editions, but at least one of the works included was originally published in the mid-sixteenth century.

The *De peditu* is a methodical discourse, indeed. Its 50 sections follow a standard rhetorical speech division: definition, etymology (from [dubious] Greek, and including Saxon and German), subcategories, causes, and effects both good (therapeutic) and bad. We learn all sorts of fascinating details, including that farts can be divided into 'vocal' and 'mute' (section 10) or into simple and multiple (15); that multiple farts, which can also be called *Diphthongi*, make a sound like *papapax, papapapax, papapapapapax* (17, quoting Aristophanes); that farting can be musical (29) and can cure illness (38–9, the Emperor Claudius again); that farts can be artificially provoked (43), and that it is a good idea to pee and fart at the same time (48). These fifty sections occupy most of the work, and are followed by seven 'questions,' which inform us, for instance, that a Pope's fart is worthier (*dignior*) than a priest's, that a fart can be chemically distilled to obtain a quintessence of farts (6), and that there are 62 different kinds of musically-sounding farts (7, quoting Cardano).

This is the most elaborate case of intellectual in-joking on a scatological topic that I have so far found, and it is tantalizing not to know who composed it, and when. Of the authors quoted in this article, I suspect all would have found it delightful, and it is certainly interesting that what the French like to think of as the 'Classical' seventeenth century, with its order, harmony and purity of language, liked it sufficiently to reprint the volume a number of times.

ꙮ ꙮ ꙮ

As I implied at the beginning of this essay, a sampling of such disparate material does not encourage any sweeping conclusions, but it does prompt some general reflections. We may deduce that quite a few German and Italian Humanists (but no French) delighted in scatological humor, that French authors of narrative used the theme of farting in very different ways, which include a surprising connection with rhetoric, and that pleasure in making gross fun of scholastic rhetoric and logic persists well into the seventeenth century. We have also learned that, besides provoking laughter in the audience or the readers, farting can be useful or profitable: it can help to cure illness, amuse and conciliate a patron, save a man from the Devil, or even (in Rabelais) engender pygmies.

Farting seems to have been more common, and more acceptable, than it is today, both in public (church, diplomatic occasions) and in private (especially in bed), but there are no obvious sociological implications in my examples. Farters come from all walks of life; we have met cardinals and court jesters, priests and magistrates, rich men, learned men, artisans and farmers. With the exception of Poggio's young wife, no women fart deliberately, whereas men seem able to do so at will. Involuntary farting is a cause of shame to women, which may explain why my examples show either middle-class or working-class women who fart: maids, servant girls, attendants, young wives, old women, 'a shy girl,' or simply 'a woman.' The most 'upper-class' woman in the group is probably Béroalde's Italian courtesan....

Undoubtedly the most surprising aspect of all this, to us as it would have been to Cicero, is the connection with rhetoric. On the scale of human preoccupations, surely rhetoric, the art of persuasion using carefully chosen language, is among the most elevated, and scatology by far the lowest and most despicable. Rhetoric typifies the artificial (in a good sense – the ability of man to improve upon nature), while bodily excretion is the most obvious form of the grossly natural of which 'cultivated' man does not wish to be reminded. So what is the point of the reminders that the two are not so far apart, when Bebel's magistrate pretends that successive farting is orderly, when Arlotto and Tabourot's storytellers develop elaborate metaphors about farting, or when the comic Discourse takes a fart as fit subject for erudite analysis? Only in Tabourot's case, I think, is the motive clear: he is systematically poking fun at the rhetorical tradition of suitable stories told by lords and ladies in an aristocratic setting.

The only firm conclusion I should like to suggest is that this is a rich, largely untapped field of research, which I hope others will pursue at greater length than I have been able to do here.

Notes

[1] This article is a revised and expanded version of a paper given at the Sixteenth Century Studies Conference in Cleveland in November of 2000. I am grateful to the organizer, Jeff Persels, and to several members of the audience for their helpful comments. Many thanks also to my colleagues Chris Brunelle and Laurie Johnson for help with translation from Latin and German (otherwise English translations are my own), and to Russell Ganim for his constructive criticism. References to joke-books include individual book numbers, where appropriate, and joke number followed by page number.

[2] See the bibliography in my *One Hundred Renaissance Jokes.*

[3] In his *Convivium fabulosum* (*The Fabulous Feast*), see *Colloquies* 266.

[4] I am quoting the bilingual edition by Marcello Ciccuto. References to jokes are to joke number and page.

[5] I am quoting the Basel edition of 1566.

[6] I am quoting a 1609 edition which includes *facetiae* from a number of other authors, including Poggio and Bebel.

[7] Florence Weinberg has argued persuasively that this apparently trivial episode has important polemical content (the Ruachians are Dutch Protestants).

[8] I am quoting from the edition by Hélène Moreau and André Tournon. For a bibliography on the book, see Michel Renaud, *Pour une lecture du* Moyen de parvenir *de Béroalde de Verville*, and my 'Il faut donner dedans: sexe ou/et rhétorique dans le *Moyen de parvenir*.'

[9] Huchon's critical edition of Rabelais explains that one player lifts up another by his feet and makes him walk on his hands, like a wheelbarrow (1122 n. 1).

[10] I quote the Slatkine Reprint (1969) of the Brussels 1866 edition.

[11] Information about Tabarin is both sketchy and confusing; it is not even clear whether his name was Jean Salomon or Antoine Girard.

[12] Frankfurt, 1615, sometimes listed under Obsopaeus. According to Brunet, this is the first edition.

Chapter 2

'The Wife Multiplies the Secret' (AaTh 1381D): Some Fortunes of an Exemplary Tale

Geoffrey R. Hope

The final story in the *Violier des histoires rommaines* (1521), the French translation of the *Gesta Romanorum* (late thirteenth, early fourteenth century), is categorized as a traditional tale called 'The Wife Multiplies the Secret' with the designation 1381D in the Aarne-Thompson (AaTh) classification. In Tubach's *Index Exemplorum* the story is number 1359, 'Black Crow from Husband's Body' and in Ranke's *Enzyklopädie des Märchens* the story is discussed under the entry 'Die geschwätzige Frau' ('The Gossipy Woman').[1] The story in *Violier* 151, 'Comment les femmes mentent souvent oultre ce qu'elles ne pevent tenir leur secret' ('How Women Often Lie Besides Which They Cannot Keep a Secret'), a translation from *Gesta Romanorum* 125, has a cleric challenge his brother about the trustworthiness of women.[2] In order to decide the argument, the married brother decides to 'le vray experimenter' ('to test the truth') by making up the following story:

> Il dist une nuit à sa femme que si elle vouloit tenir son cas secret, qu'il luy diroit merveilles. Mais au contraire, qu'elle le feroit confuz et infame.
> – Ne crains point, dist la femme, car tu scez bien que toy et moy ne sommes q'un corps. Jamais ton secret ne revelleray.
> Parquoy il luy dist qu'en allant à son secret de nature, luy estoit de la partie posterieuse sailly ung corbeau noir com ung dyable, dont il estoit dolent. (*Violier* 457)[3]
>
> He said one night to his wife that if she wanted to keep his case secret, that he would tell her marvels. But if not, that she would have him confounded and lose his good name. — Do not fear, said the wife, for you know well that you and I are but one body. I will never reveal your secret. So he told her that while going to his secret of nature, from his posterior parts came out a crow as black as a Devil, which upset him.

Despite her promise, the wife proceeds to tell her neighbor how two crows had flown out from her husband. That neighbor then goes even further telling someone else that it was three crows. The story goes around the neighborhood and eventually it tells of 40 crows (60 in the *Gesta*) at which point the husband calls the people together, tells the truth of how he had experimented on, or tested, the lies of women ('comment il avait experimenté le mensonge des femmes' [457]; in the *Gesta*: *quomodo mulierem voluit experiri, si sciret secretum tenere* [476], 'how he meant to put women to the test, [to see] if they were able to keep a secret').

Then, in a narrative shorthand that is typical of the *Violier* and of narrative *exempla*, the wife dies and the widower becomes a monk. The story, as is standard in this kind of *exemplum*, is then put into a Christian moralization in which elements of the anecdote are allegorized to demonstrate symbolic relationships between the soul and the world. Here, the husband is the worldly man who falls into sin, symbolized by 'saillir et voller du derriere le noir corbeau' (the black crow issuing forth and flying out of the bottom) (458). It is characteristic of the relationship between the tale and the moralization in this book that the husband who plays a more or less positive role in the tale (by unmasking falsehood), is the sinner in the moralization. The purpose of the book is to demonstrate that the world may appear valuable but that it must be abandoned for the rewards of heaven.

This story takes meaning in a number of contexts. As I have shown elsewhere (Hope 1997), the anonymous translator of the Latin *Gesta* moved this story so that when *Gesta* 125 (out of 181 tales in the *Vulgärtext*, cf. Oesterley) is turned into French and becomes *Violier* 151 it occupies the last place in the collection. An important aspect of the *exemplum* genre as typified in the *Gesta* and the *Violier* is the constant move from the anecdotal part of the *exemplum*, the story or tale, to the clerical moralization in which values of the tale are inverted in order to be rejected along with the world. *Violier* 151 exemplifies this rejection of the world to which is added the rejection of the book of stories itself through the image in this newly final story of a fiction that turns into lying gossip in the expanding story of excremental birds. The story within the story shares with the other tales of the collection a particular exemplary function: it is told as a made-up story, though one that purports to be true, in order to determine and demonstrate a larger truth. The image of the tales not only as false but also as ever-multiplying feces is effective as a closural device in the *Violier* as it merges with the image of the world which is, just like the story, something awful that is to be stopped. This book of stories ends with a sort of purge of stories.

There are many versions of the tale type AaTh 1381D. The motif that women cannot keep a secret may well have been traditional at the time of Plutarch and of Aulus Gellius, both of whom tell non-scatological versions of the story as an *exemplum* to warn about the dangers of talking too much and, in particular, about the inability of women to keep a secret.[4] Macrobius (1.6) tells the story following Gellius, and that version is the one told in the *Exempla* of Jacques de Vitry (number 235 in the Crane edition). This is also the version presented in the chapter of the *Gesta Romanorum* that directly follows 'Black Crow from Husband's Body' (V151), Oesterley 126, *Quod mulieribus in nullo est credendum et presertim de secretis celandis. Exemplum jocundissimum* (476–77; 'That Women Are Not to Be Believed and Particularly in the Keeping of Secrets. A Most

Delightful Example'). Tubach classifies that story as number 5269, 'Wife and Secret Senate' (motif J1546 in Stith Thompson). Erasmus (*Opera omnia* 4: 1.233–370) recounts the story following Plutarch in his *De lingua* and it is from these sources that Noël du Fail picks up the anecdote in *Contes et discours d'Eutrapel* (2: 311–12). Rabelais tells a variation of the story in the *Tiers Livre*. These versions of the story are as follows. According to Plutarch (7: 1.219–57), the Roman Senate was working late on a particularly difficult problem (we never learn what that problem is) and postponed the decision to a later date with the understanding that, until that time, the senators should remain silent about the matter. One senator's wife (in Gellius it is the mother of a boy who had been permitted to observe the Senate's deliberations) asks what the problem was. The initial refusal to tell the woman what is going on is countered by whatever means a woman has at her disposal: the wife uses tears and blandishments; the mother adds a threat to beat the boy. And so the husband (or son) makes up a fantastic tale and elicits the woman's promise not to reveal the secret. In Plutarch, the made-up story is that a bird has been noticed flying over Rome dressed in a helmet and carrying a spear and that the augurs and diviners have not made up their mind as to the significance of this sign, be it war or peace. In Gellius (1.71–2), the tale is that the Senate has been deliberating as to whether men should be allowed to marry two women or women to marry two men. In both versions, the woman then breaks her promise and recounts the story to a neighbor, saying, in Plutarch's version of the story, that not one but a number of such birds have been seen. The word soon becomes general. In Plutarch, the senators are confused when they start hearing the story from their wives that hundreds of armed birds have been seen in the sky. In Gellius, the wives in Rome confuse the senators by presenting a petition in favor of women marrying two husbands and opposed to allowing men to marry two wives. The original senator (or the son) recounts his deception and the lesson is clear to all.

In these tales, a story within a story is told that puts a woman's ability to keep a secret to the test – a test that she fails. These stories may be said to develop the theme of fiction as a means to determine the truth, i.e., that women are not to be entrusted with a secret. The anti-feminist tenor of this Classical *exemplum* is political in nature: the place of women is not to be in the Senate of Rome. The context in Aulus Gellius, Plutarch, and in Erasmus and du Fail, however, is only incidentally anti-feminist, as in all of these cases the *exemplum* is one of many told about the dangers of talking too much in general. That women are what they are and do what they do simply helps illustrate the concerns of talkativeness in society and particularly in public policy. In these versions of the tale, the secret given to the wife is also of a political nature. The secret is of course nothing but a fantastic tale of anthropomorphic birds, or of a simple but radical change in the marriage laws, but the significance of both secrets is political in that matters of war or peace, and the establishment of, or changing of laws, is supposed to be judged appropriately by the Senators, not by women and their neighbors.

Rabelais has Ponocrates tell a version of this anecdote in chapter 34 of the *Tiers Livre*. He presumably picked the story up from an anti-feminist poem by Gratien du Pont, the 1534 *Controverses des sexes masculin et feminin*. He may also have known the story from sermon literature: it is found as an *exemplum* at the end of sermon 50 of the *Sermones discipuli de tempore et de sanctis* (1474) by Johann Herolt. The story is prefaced in *Tiers Livre*, chapter 33 when Hippothadée refers to a learned tradition according to which 'la premiere femme du monde, que les Hebrieux noment Eve, à poine eust jamais entré en tentation de manger le fruict de tout sçavoir s'il ne luy eust esté defendu' ('the world's first woman, whom the Hebrews call Eve, would hardly at all have entered into the temptation to eat the fruit of all knowledge if it had not been forbidden to her') (*Œuvres* 458). In a variation of Genesis 3, Hippothadée has the Devil, 'le Tentateur cauteleux' ('the devious tempter'), establish the necessity of eating the fruit by the very fact that she is a woman: 'Il t'est defendu, tu en doibs doncques manger: ou tu ne serois pas femme' ('It is forbidden to you; therefore you must eat some of it or you would not be a woman') (458). This fanciful reference leads to chapter 34 called 'Comment les femmes ordinairement appetent choses defendues' ('How women usually have an appetite for forbidden things') which includes as an example of that principle, the story of the good sisters of Coignaufond.[5] Here, the 'abesse et meres discretes' of the abbey ask Pope John XXII for permission for them to hear their own confessions: 'alleguantes que les femmes de religion ont quelques petites imperfections secretes, les quelles honte insupportable leurs est deceler aux homes confesseurs' ('alleging that religious women have a few small, secret imperfections which it is an unsupportable shame to reveal to male confessors') (459). The pope questions their ability to keep a secret at all which they affirm: 'et plus que ne font les homes' ('better than men do') (459). Promising to return the next day, the pope leaves them a closed box which has a bird in it and enjoins the 'bonnes dames' not to open the box 'sus poine de censure ecclesiasticque et de excommunication eternelle' ('on pain of ecclesiastical censure and of eternal excommunication') (459). They open the box as soon as he is gone, and when he returns to grant their request, he finds the bird gone, proof that they would not be able to keep the secrets of the confessional. Rabelais's story may recall Pandora but the context of the church confirms a relationship with the Classical versions outlined above: because they are unable to control themselves and to keep a secret, women are not suited for the professional responsibilities that are associated with a major institution.

There is another group of analogs of the tale, a group that appears to come from eastern sources, in which the context is not political. The story recounted in *Violier* 151 is associated with this tradition. In 'The History of the Forty Vezirs' (a Turkish version of the well-known eastern frame-tale 'The Seven Wise Masters') as translated by E.J.W. Gibb (1886), the tale is of a farmer who discovers a wonderful amount of treasure while plowing his field.[6] He is hesitant to tell his wife about it, however, because he is not sure she will keep the secret. To test her, he invents the story that, while plowing, a crow flew out from him, which he tells her only when she swears secrecy. The wife tells the tale to a neighbor, exaggerating it so that the story is about ten crows and soon the story comes back to the husband concerning a hundred crows and the husband has the answer to his question about his wife's ability to keep a secret. When he learns he cannot trust his wife, he tries other means of arranging affairs to be able to keep his treasure but nothing works and he ends up knowing where the gold is but not being able to enjoy it. This story is close to *exemplum* 125 of the *Gesta Romanorum* (*Violier* 151). In the *Gesta*, there is no treasure to hide, only the question of a wife's ability to keep a secret and not to lie. In *The Exempla of the Rabbis*, Moses Gaster presents an analog to this tale (*exemplum* 56) in early Jewish traditions: 'Woman cannot keep a secret. A Hegemon twits a Jew with that saying, and is convinced of the truth by feigning illness and telling his wife under an oath, that he had become pregnant with child through the king. She divulges it nevertheless though he said he might lose his life through it' (65 and 196). What appears to be a variation of this tale appears in fable 129 of the *Hecatomythium* of the Italian Humanist Laurentius Abstemius (1495), *De Viro qui Uxori se ovum peperisse dixerat* ('About the Man Who Told His Wife He Laid an Egg') in which, as in the *exemplum* from the *Gesta*, there is no other motive to the story than simply to test the wife, only this time the story is that the husband laid an egg.[7] This story is the one La Fontaine tells in fable 6 of book 8, *Les Femmes et le secret*.[8]

While actual sources and derivations would be interesting, that kind of relationship among versions of the story will be difficult to fix because of the oral history of the tale type which clearly exists in folklore traditions. In the absence of that kind of knowledge, it is nonetheless possible to suggest that the two groups of stories presented above show significant differences and that these variations may help account for the different paths the stories have taken in literary history. The stories of Plutarch, Aulus Gellius, Macrobius, *Gesta* 126, Rabelais, Erasmus and du Fail may be referred to as the 'Classical' group. The other group may be identified by its apparent origins as the 'eastern' group, and includes the story from the *Forty Vezirs*, the *exemplum* of the rabbis, *Gesta* 125 (*Violier* 151), and the fables of Abstemius and La Fontaine. The groups are linked by the test of keeping a secret that a man gives a woman and that the woman fails. That is, both groups develop the theme of fiction as a device, as a tool in the plot, and both develop an anti-feminist message through the woman's immediate inability to keep her word as well as through the participation of other women in the way in which she fails the test.

One tale type finds illustration in two very different traditions. Differences between these groups relate to narrative, thematic and generic frameworks. These differences help reveal how a story can either remain a conservative demonstration of a cultural understanding of the world or quite radically undermine the basis for that understanding.

One of the differences between the two groups involves the nature and the resonance of the conflict. In the story Plutarch tells, the problem posed and illustrated may be seen to be a stable – that is, an ongoing and permanent – conflict between men and women concerning a public institution, the Roman Senate. The same is true in Gellius. In Rabelais, the Senate is replaced by the Church. These stories illustrate why a woman's place is not to be in the public arena. At the conclusion of these stories, no change to the initial situation has occurred. By their actions, women have simply demonstrated their nature. In story 151 from the

Violier, on the other hand, the conflict is developed less between man and woman in general than between husband and wife. The conflict does not involve a separate, professional matter but puts the focus on what binds the two people together: the institution of marriage. The nature of the secret here, as in the *Forty Vezirs* story, in Abstemius, La Fontaine, and the *exemplum* of the rabbis, is shameful for the husband. By telling this story the women here are not simply betraying their word, they are also revealing an embarrassing event concerning their husbands, a revelation that is presumed to have consequences. While none of these stories continues to show directly the effect of this betrayal on the marriage, *Violier* 151 does point broadly to such an effect: 'Puis sa femme mourut, et il se fit moine' ('Then his wife died, and he became a monk') (457). The wife's death is not explained but may easily be interpreted as a symbolic divorce: the two cannot live together any longer. Similarly, in the Jewish tale, revealing the secret is said to put the husband's life in jeopardy.

Stories in both Classical and eastern groups share exemplary status and are told in order to illustrate a cultural concept, a moral or some sort of truth. It is also possible to suggest a difference between the two groups in this regard. The only example I have found of the Classical group that is presented as an independent story rather than as an example within a larger discourse is *Gesta Romanorum* 126, 'Wife and Secret Senate.' The *Gesta* presents *exempla* and other stories in ways that move them from the more instrumental functions of sermon traditions, as memory aids and rhetorical illustration for preachers. Additionally, they give a context in which they are read for themselves, thereby providing something akin to a modern understanding of literature.[9] By being translated into French as *Violier* 151, *Gesta* 125 acquires a new public, one that would be even more secular and non-professional than the readership of the *Gesta*. By virtue of its significant place at the end of the *Violier*, that scatological story furthermore helps give a new unity to the collection of stories. *Gesta* 126, however – a quite similar story – is one of 30 from the *Gesta* that are not included in the *Violier*. One way of understanding the choice to include one and not the other in the French book is that *Gesta* 126 is perceived as more of an example, an *exemplum*, less apt to figure in a literary context, than the former. The other Classical versions are all told as examples to illustrate a point. They support meanings from the larger text within which they are included and do not have, are not given, independent existence as stories. They serve with a number of other such anecdotes to demonstrate the dangers of talking too much, the impossibility of trusting women, and, in particular, the unsuitability of women for public office. The eastern versions do share a kind of exemplary status with the Classical versions but they tend at the same time to establish themselves as autonomous texts, to exist and to be enjoyed and profited from as such rather than simply as aspects of a larger, discursive text on a topic such as *Garrulitate* (Plutarch), *Lingua* (Erasmus), or the nature of women (Rabelais). It should be noted that this distinction between independent story and included example illustrates what I perceive as a literary-historical tendency, certainly not a structural difference between types. In Juan de Mal Lara's *Filosofía Vulgar* (1568), Abstemius's fable is told as an example within a larger discussion, not as an independent story (1: 387–88).

The relative generic independence of the eastern versions may be seen in the same light as their more personal, less institutional, though perhaps no less ideological, thematic framework. The test the woman fails in these stories has personal significance in terms of the relationship between husband and wife, and social significance in terms both of the relationship among women, based on gossip, and on the position, or status, of man as projected by women. Part of the social significance I am referring to here is shared by both groups of the tale: the woman does not act alone but always in conjunction with other women. Even in Rabelais's story, which is not about gossip, the inability to keep a secret is a shared female trait. What distinguishes the eastern versions is precisely that the secret revealed concerns the bodily lower stratum: a fantastic defecation, the laying of an egg. It would appear to be this motif that gives these versions the chance to develop as relatively independent narratives: an interest not only in women's willful ways but also in the personal as a sign and token of privacy, of an intimate contract that the husband assumes, and that the wife affirms (though she is lying), to be part of marriage.

In the *Violier*, the husband asks his wife to 'tenir son cas secret' ('to keep his case secret'). The term he uses to designate his bowel movement is 'secret de nature' ('secret of nature'). The wife affirms that she will keep the secret on the basis of their marriage, because 'toy et moy ne sommes q'un corps' ('you and I are but one body') (457). This reference to Genesis 2.24, *et erunt duo in carne una* ('and they shall be one flesh'), may be understood as an ironic reaffirmation of her commitment to the marriage and so makes it explicit that she is betraying their marriage and his body as it is joined to her in that marriage. There is a sense of embarrassment in that she is broadcasting her husband's bowel movement, his 'secret de nature,' his bodily function, to the whole world at the same time as she is demonstrating no allegiance to the marriage bond.

Both types of the story present not only a view on women but also a perspective on man. In Plutarch and related stories the woman initiates the request for a story, while in stories like *Violier* 151 it is the man who decides to give a test. In the former, it is shown that, by their actions, men merit power and that women do not; in the latter, men tend to find themselves defined in one way or another as deformed, strange, and as outside of the community. In the Classical types, men affirm and justify sole control over a public but exclusively male institution. In the others, men lose control, in the context of a much more general kind of public, over the definition of their bodies.

It is possible that the story of the husband laying an egg in Abstemius and later in La Fontaine is a transformation of the crows as feces image in the direction, away from scatology, of some neo-Classical notion of *bienséance*. It would appear, however, that this transformation is concerned less with cleaning the story up than in building up the stakes of the test and the significance of its failure. There is an intensification of the shamefulness of the story about the egg in comparison to the crow version. In La Fontaine, the husband's injunction of silence is explicit regarding the dangers of the story being known: 'On m'appellerait poule' ('I would be called hen') (9).[10] The husband here is not simply part of a fantastic event involving his lower parts, he is both animalized and feminized by the story he tells. His body is revealed in the same way as in the *Gesta* version, and it similarly undergoes a fantastic sort of transformation, but it is also made into the body of a female bird. The betrayal here is of the marriage and the body but also of the man's masculinity and humanity.

The point that may be made in conclusion involves a changing relationship, observed in these various tales, between men, women, and the public world. The stories that follow Plutarch and Gellius present what might be referred to as a kind of Classical anti-feminism: women talk too much, they cannot keep secrets, and are thus not suited for public life. The stories in the eastern group present a different kind; a more narrative anti-feminism in which women are presented more as characters in fiction where they play important and destructive roles. Where women are the butt of the joke in Plutarch, Gellius, and Rabelais, in *Violier* 151 and the stories like it, women are gullible liars but it is the man who, by taking the initiative of the test, ends up losing control and definition, as it were, over the image of his body. While Classical versions purport to keep women in their place, these stories recount the dangers, and perhaps the impossibility, of domesticity.

Henri Régnier in his 1889 edition of La Fontaine's *Fables* helps confirm an anxiety that is brought out in the eastern versions when he writes concerning 'Les femmes et le secret': 'Il faut peut-être regretter que la Fontaine n'ait pas préféré aux données du fabliau d'Abstemius celles d'un récit de Plutarque, dont l'objet est le même, qui ne le cède pas à l'autre en gaieté, et qui est d'une invention plus naturelle à la fois et plus élégante' ('It is perhaps necessary to regret that La Fontaine did not prefer to the material of the fabliau of Abstemius the matter of a story in Plutarch, whose object is identical, which does not cede anything to the other in gaiety, and which is of an invention that is both more natural and more elegant') (238). It is not difficult to read Régnier's concerns and preferences as typical of a male-dominated culture steeped in Classical example. That women maintain their own public arena, through gossip, and have no aptitude for the public and professional arenas of men, makes no challenge to traditional cultural assumptions. If, on the other hand, women's public voice broadcasts information about men, information that is as private as it is false – and manages to change, to deform and degrade – the public image of man, that may certainly be perceived not simply as inelegant but as against some sort of nature.

Notes

[1] The entry in the *Enzyklopädie des Märchens* develops five groups of tales under the larger complex 'The Foolish Wife and Her Husband' (AaTh 1380–1404). The stories I am discussing are for the most part included within the fifth of those groups.

[2] Oesterley, ed., *Mulieres non solum pandunt secreta, sed ad hec mentiuntur plura* (*Of Women, Who Not Only Betray Secrets, But Many Also Lie*) (475–76). All translations, except as otherwise noted, are my own.

[3] *Carissima, secretum habeo; tibi panderem, si certus esse, quod nulli dices; quod si contrarium facies, confusio intolerabilis michi esset. At illa: Domine, noli timere, unum corpus sumus; bonum tuum est meum et e converso et eciam malum similiter. Qui ait: Cum ad privatam accessissem, ut opus nature facerem, corvus nigerrimus a parte posteriori evolabat, de quo sum contristatus.* (My dear one, I have a secret; I would reveal it to you if it was certain that you would tell no one; because if you do the opposite, it would be an impossible confusion for me. But she says: Lord, do not fear, we are one body; your good is my own and conversely [your] troubles are the same [as mine]. So he says: As I went to the privy, to do a natural function, a very black crow flew out from my posterior parts, which makes me sad) (Oesterley 475). For the *Violier* text, see my edition. On the *Gesta Romanorum* see, besides Oesterley, B. Weiske and the entry in the *Enzyklopädie des Märchens*.

[4] In Stith Thompson's *Motif-Index of Folk Literature* 'the Wife multiplies the secret' is motif J2353.1. This motif is included in Rotunda's *Motif-Index of the Italian Novella in Prose*. Other related motifs include H1569.4 and X39, both of which are included in Goldberg's *Motif-Index of medieval Spanish Folk Narratives*.

[5] Donald Frame (360) translates *Coingnaufond* as 'Thrustitindeep.'

[6] See Chauvin 8.168. See also the articles 'Sieben weise Meister and Vierzig Wesire' (when they appear) in the *Enzyklopädie des Märchens*. Two versions of this book that I have not been able to consult are Belletête (which was used by Gibb), *Contes Turcs en langue turque* (1812) and Behrnauer, *Die Vierzig Veziere oder Weisen Meister* (1851).

[7] On Abstemius and the *Hecatomythium* see the article in the *Enzyklopädie des Märchens* and the chapter entitled 'Les "Fables" et la tradition humaniste de l'apologue ésopique' by Marc Fumaroli in his edition of the *Fables* of La Fontaine (lxxxi–ciii). Pierre Bayle also has an entry on the author in the *Dictionnaire historique et critique* (1697).

[8] H. Régnier, in his edition of La Fontaine's *Fables* (2.238–41), quotes Lodovico Guicciardini, *Detti et fatti*: 'Cosa stolta et pericolosa communicar alle donne segreti importanti' (A stupid and dangerous thing to communicate important secrets to women). Also see the notes by Marc Fumaroli in his edition of the *Fables* (889-90).

[9] Claude Bremond writes : 'la thématique narrative et didactique condensée dans les recueils d'exempla, et la forme compilatoire elle-même, peuvent échapper à leur fonction initiale d'aide-mémoire et évoluer vers des compositions dont la lecture n'est plus instrumentale, mais devient une fin en soi (ainsi le Traité d'Étienne de Bourbon, les allégorisations de Jean Gobi ou des *Gesta Romanorum*...)' (the condensed thematic and didactic narrative in collections of *exempla*, and the compilation form itself, can escape from their initial function of memory aid and evolve towards compositions in which the reading is no longer instrumental but

becomes an end in itself [as in the Treatise of Étienne de Bourbon, the allegorizations of Jean Gobi or the *Gesta Romanorum*...]) (27–8).

[10] Régnier quotes Abstemius 129: '*Sed cave, si me amas, ne cui hoc dicas: tute enim nosti quanto dedecori mihi esset, si ex viro gallina factus dicerer.* – L. Guicciardini traduit ainsi: "Ma guarda, ben mio, se tu mi ami, che non ti uscisse di bocca, perchè tu puoi pensare che dishonore mi sarebbe se si dicesse che d'huomo io fussi diventato una gallina"' ('"But watch out, if you love me, not to say this to anyone: for you know how great a disgrace it would be to me, if it was said I was turned from a man into a hen." L. Guicciardini translates thus: "But be careful, my dear one if you love me, that it does not come out of your mouth, because you can imagine what a dishonor it would be to me if it was said that from a man I had turned into a hen"') (239).

Chapter 3

Doctor Rabelais and the Medicine of Scatology

David LaGuardia

Rabelais describes the beginning of Gargantua's adolescence and education from two points of view. The first is his tutelage at the hands of 'Sophist' preceptors, the second is his more formal education under the direction of Ponocrates, the ideal Humanist pedagogue. The first of these procedures is prefaced by that most scatological of Rabelaisian chapters, in which Grandgousier recognizes his young son's superior intelligence thanks to the latter's discovery of the most voluptuous manner of wiping his backside. As Jeffery Persels has demonstrated, there can be no doubt as to the significance of this childish diarrhea: Gargantua has not yet been properly educated, and the tons of drivel that enter his mind every day can only be transformed into the mountains of excrement that he expels in inappropriate places ('Straightened' 107–8). As Persels has pointed out, these scatological passages have to be interpreted in theological terms: the babble of a corrupt faith enters the mind and is expelled from the body as foul, unhealthy excrement, just as the excessive food that one eats in an undisciplined way causes the body to be sick and to expel noxious refuse. He also cites the example of the debates between Luther and Thomas More, in which each of these great men accused the other essentially of wallowing in doctrinal shit. In this scatological register, constipation is thus equal to theological confusion and error, while the healthy 'excrétion des digestions naturelles' in an appropriate place demonstrates a control of the sphincters that is the material equivalent of a proper spiritual understanding of the divine (103–6).[1]

In contrast, the first stage of Gargantua's Humanist education is a purging of the impure elements that have poisoned his body:

> Pour doncques mieulx son œuvre commencer, [Ponocrates] supplia un sçavant medicin de celluy temps, nommé maistre Theodore, à ce qu'il considerast si possible estoit remettre Gargantua en meilleure voye. Lequel le purgea canonicquement avec Elebore de Anticyre, et par ce medicament luy nettoya toute l'alteration et perverse habitude du cerveau. (64)

> So in order to begin his task better, [Ponocrates] requested a learned doctor of that time named Master Theodor, to consider whether it was possible to put Gargantua back on a better track; and he purged him canonically with

> hellebore of Anticyra, and by that medicament scoured out all the alteration and perverse habit of his brain. (55)

Scholars are agreed that one must read these concrete and visceral passages of Rabelais in the most abstract of terms: the physical purgation of the body is an exact analog of the spiritual purgation of the soul. One need go no further than the preceding passage in order to justify this 'spiritual' interpretation of a grossly physical act that was often considered a medical necessity throughout the Middle Ages and the Renaissance: the purgation of the lower bowels literally purifies the functioning of the brain here, and helps the individual to digest and to expel the excess humors that impede the functioning of his body. Moreover, an ingenious pun that was highlighted by Guy Demerson in his edition of the text supports this reading. 'La *Purgation canonique* (expiation légale) était un chapitre du recueil de droit, le *Digeste*' ('*Canonical Purgation* [legal expiation] was a chapter of the legal compendium, the *Digest*') (106 n.3). Typically, Rabelais's text works here on several distinct levels at the same time. The literal purgation of the giant's body is also a kind of juridical purification. One must also read the spiritual connotations of this legal gloss, however, in terms of the standard metaphor that links thinking and learning to the stomach and digestion. If, as Demerson suggests, this passage is indeed a reference to the *Digest of Justinian*, the major compendium of Roman and Byzantine law for the Middle Ages, the very title of this reference unavoidably leads us back to the digestive tract. One can only marvel at how Rabelais transforms what one would assume to be a simple enema into a micro allegory of the purgation and purification of the soul, as well as the beginning of the legal incorporation of an unruly adolescent into the adult world.

In contrast to the scatological yet spiritual renewal that the giant experiences at the hands of his Humanist tutors, his typical day following the guidance of his 'Sophist' preceptors begins in a drastically different fashion:

> Puis [Gargantua] se guambayoit, penadoit, et paillardoit parmy le lict quelque temps, pour mieulx esbaudir ses esperitz animaulx....
>
> Puis fiantoit, pissoyt, rendoyt sa gorge, rottoit, pettoyt, baisloyt, crachoyt, toussoyt, sangloutoyt, esternuoit et se morvoyt en archidiacre, et desjeunoyt pour abatre la rouzée et maulvais aer : belles tripes frites, belles charbonnades, beaulx jambons, belles cabirotades, et forces soupes de prime. (56)

> [Gargantua] waggled his legs as he sat, bounced about and tumbled in the bed for some time the better to brighten up his animal spirits....
>
> Then he crapped, pissed, threw up, belched, yawned, farted, spat, coughed, sobbed, sneezed, blew his nose like an archdeacon, and ate breakfast to put down the bad air: fine fried tripes, beautiful carbonadoes, fair hams, fine game stews, and many early morning dips as snacks. (49)

In this highly overdetermined passage, the catalog of exaggerated bodily eruptions is followed by its somehow logical reversal, the ingestion of immense quantities of fatty meats, which medieval and Renaissance doctors considered a fundamental yet potentially deleterious element of one's diet. Almost every medical text on health and hygiene from the period contains a catalog of the properties and nutritive values of different types of animal and bird flesh. Beef, for example, was considered to be cold and dry, and to give rise to thick blood and melancholy, at least according to a thirteenth-century text, Aldebrandin de Sienne's *Le Régime du corps* (*The Regimen of the Body*), which catalogs the digestive effects of all kinds of meat, from goat to rabbit to duck and partridge, and of many types of organs and body parts, such as poorly nutritive ears and noses to feet and tails, which the text says 'produce abomination.'[2] All of these medical and hygienic texts describe the dangers of eating too much meat, which they believed could literally have deleterious effects on the brain, as we will see below.

The contrast between Gargantua's uncontrolled dietary habits and those he assumes after his canonical purgation is quite remarkable. As a consequence, his excrement becomes more 'rational,' and becomes part of his normal daily routine early in the morning, following the guidance of his Humanist tutors:

> Se esveilloit doncques Gargantua environ quatre heures du matin. Ce pendent qu'on le frotoit, luy estoit leue quelque pagine de la divine escripture haultement et clerement avec pronunciation competente à la matiere, et à ce estoit commis un jeune paige natif de Basché, nommé Anagnostes. Selon le propos et argument de ceste leçon, souventesfoys se adonnoit à reverer, adorer, prier et supplier le bon Dieu : duquel la lecture monstroit la majesté et jugemens merveilleux.
>
> Puis alloit es lieux secretz faire excretion des digestions naturelles. Là son precepteur repetoit ce que avoit esté leu : luy exposant les poinctz plus obscurs et difficiles. (65)

> So Gargantua woke up about four o'clock in the morning. While they were rubbing him down, there was read to him loud and clear some page of the Holy Scripture, with a delivery appropriate to the matter, and assigned to that was a young page, a native of Basché, named Anagnostes. According to the gist and argument of this lesson, he often gave himself up to revering, adoring, praying, and beseeching the good God, whose majesty and marvelous judgments the reading demonstrated.
>
> Then he went to the private places to make an excretion of natural digestions. There his preceptor repeated what had been read, expounding to him the most obscure and difficult points. (55)

Demerson remarks that, 'Cette discrétion s'oppose à la volubilité scatologique du chapitre 13; mais cette rumination, en un tel lieu, des matières les plus difficiles à ingurgiter, n'est sans doute pas présentée sans ironie' ('This discretion contrasts with the scatological volubility of chapter 13; but this reflection, in such a place, upon the most difficult materials to be ingested, is hardly presented without irony') (107 n.10). While Rabelais could rarely resist an opportunity of mixing scatological irony with the most serious of matters, such as the reading of Holy Scripture, it is clear here that the natural excretion of that which the giant has digested goes hand in hand with the metaphorical digestion of sacred discourse. A Humanist education, which included a careful reading of Scripture, was food for the mind, with a subsequent 'passing on' of that food to the end of the digestive tract. This allegorical sense of these passages is more than evident, and has been explicated by numerous scholars in diverse contexts. A healthy regimen of reading was thus quite akin to, and literally the same as, the health and proper functioning of the body in this context, which of course depended upon the orderly digestion and expulsion of nourishment.

The religious and theological context for Rabelais's seemingly scandalous preoccupation with scatology has, therefore, been well documented in the scholarly literature on the subject. What has not been examined is the notion of purgation as it is defined in the treatises that constituted the medical knowledge of Rabelais's day. It is evident that if constipation and indigestion were analogous to theological error, purgation was akin to religious purification. It is not so clear, however, what the real consequences and implications of purgation were perceived to be from a scientific point of view at the time. Was it possible for writers of the sixteenth century to think of forced scatological evacuation from a purely 'mechanical' point of view, in the sense that we might think of it today, especially given the rich and complex analogies of body and spirit that permeate the writings of the period, and which transform even an enema into an allegory for the life of the mind and spirit? In other words, for us the body is essentially a machine that functions according to rational principles, if one discounts the 'psychological' dimension of physical health. In contrast, in the passages cited above from *Gargantua*, the protagonist does not only defecate improperly during his Sophistic education because he eats too much fat, drinks too much wine, and exercises by rolling around on the bed; rather, the fact that he eats too much meat is linked to the impurity of his spirit, which must undergo a canonical purgation and expiation of its corporeal sins. According to Michel Jeanneret, the idea that the control of personal hygiene and of one's diet helped to form a person's moral character was a commonplace of Humanist thinking that the Renaissance borrowed from the ancient world.[3] From a medical point of view, however, was there a conception at this time that the scatological expulsions of the body were somehow linked 'mechanically' to the soul and the mind? If so, what was the significance of purgation from the point of view of individual health?

The important question here, then, is to know exactly what Rabelais meant and understood by the purgation to which Gargantua was subjected at the beginning of his Humanist education, resulting in a more regular excretion of the natural products of his digestion, as the text says ironically. An attentive glance at medical texts both from the ancient world, which Rabelais clearly knew and read, and from early Renaissance Europe, reveals that the mental and physical health of the body in this context was thought to depend upon the movement, mechanics, and temperatures of the four humors: red bile or choler, black bile or melancholy, phlegm, and blood. For general health, the ingestion and excretion of substances had to be strictly regulated in order to avoid a repletion of the major organs, which needed outlets for the humors that naturally issued from them. From this point of view, the working of the brain as the seat of reason depended upon the proper movement of the humors from this organ to its natural outlets. This meant ultimately that reason itself could be incapacitated if the stomach or the bowels were not doing their job properly.

This strangely holistic conception of the body is evident in Galen's *De sanitate tuenda*, which was quoted by all of the treatises on health and hygiene that were written and published from the fourteenth to the sixteenth centuries. Galen writes:

> Those, however, who are in continuous pain on account of the sensitivity of the nerves which, originating from the head, for the most part invest and supply the esophagus and extend thence to the rest of the stomach, it is the function of the hygienic art to treat, but rather to forestall completely the occurrence of this situation by the best prophylaxis, by preventing the biliary fluid from flowing down into the intestine at all, or by evacuating it as quickly as possible.... But if this is not done, the ichors flow down into the stomach, which the body collects as excrements. And from the effluvia of these, not only headaches occur in some, but also the symptoms of effusions, and in some also epileptic convulsions. (262)

Elsewhere Galen states that the normal outlets for the 'biliary fluid' excreted by the brain are the ears, nose, palate, and sometimes the eyes. If there is an excess of this fluid, however, due to physical or dietary overabundance, it seeks other outlets, often accumulating in the stomach and the lower bowels. In turn, this blockage may have a negative effect on the brain, as Galen continues: '... the phlegmatic fluids, retained in the stomach so that they erode it, send up to the head vapors which are sometimes deleterious' (263). The cure for these noxious vapors is the purgation of the stomach by one means or another, usually through vomiting or the emptying of the bowels. In some medical texts, such as the *Régime du corps*, there are entire chapters devoted to the virtues of vomiting and the best times of the year for blood letting and purgation, which are, of course, Fall and Spring. It should be noted that in this medieval text, the chapter on purgation does not describe what we would consider an enema; rather, the author gives recipes for different types of herbal and medicinal broths that patients replete with the diverse humors were supposed to drink, depending upon the constitution of the patient, and the climate in which he or she lived. For the purgation of melancholy, for example, Aldebrandin recommends a hot broth containing oxymel, which was a mixture of honey and vinegar considered to be a laxative, seneca root, which was also used for snake bite, epithyme, fennel root, anise, parsley, and so on (52). Furthermore, the way in which the doctor could judge whether or not a body that was replete with melancholy had been purged was by examining the patient's urine, and not his feces. Perhaps we need to reconsider, then, whether Gargantua's canonical purgation was scatological in the strict sense at all, as we have been assuming here all along. Moreover, in the brief section of Ambroise Paré's text devoted to purgation, the famous surgeon recommends the taking of a mixture containing hellebore and antimony as treatment for symptoms of the plague, while purgation was meant to be accomplished here through the normal digestion of these elements, mixed together and taken as pills (413–18).[4] This latter type of purgation is described in the *Pantagruel* in a parodic fashion, as we will see in a moment.

A brief examination of these medical texts demonstrates that, for late medieval and Renaissance doctors, there was a definite connection between the brain as the seat of reason and the flatulence produced by overabundant humors. One of the more interesting descriptions of the connection between the brain and the bowels is found in an English manual entitled *The Castel of Helthe*, first published in 1539 by Rabelais's contemporary Sir Thomas Elyot. In the third book of this text, Elyot speaks of the phenomenon of bodily repletion, which the ancient doctors attributed notably to an excessive ingestion of meat. Here we recall in passing the 'belles tripes frites, belles charbonnades, beaulx jambons, belles cabirotades et forces soupes de prime' (56) ('fine fried tripes, beautiful carbonadoes, fair hams, fine game stews, and many early morning dips as snacks' [49]) that Gargantua had for breakfast while under the tutelage of his Sophist preceptors. The consequences of this carnivorous excess can only be what Elyot calls a 'superfluous abundaunce of humours in the body' (51), resulting in Galen's 'deleterious effects' on the brain itself. Elyot continues:

> ... the meate that shall make sychenes, must not a lyttell excede the exquisite measure.... And of that, whyche is resolved of meate undygested, procedeth fumosytie, grosse and undygested, which ascendynge up into the head, and touchynge the ryme, wherin the brayne is wrapped, causeth head ache, trembling of the members, duskynes of the sight, and many other sycknesses: also by the sharpnes therof, it prycheth and anoyeth the sinewes, which make sensibilitie, the rootes of whom are in the braine and from thens passeth throughe all the bodye. Finallye, the sayd fumositie, ingendred of replecion, percynge the innermooste parte of the sayd synewes, called sensible, it grevousely annoyeth the power animall, there consistinge, by the occasion wherof, understandynge and reason, as to the use of them, are lefte and troubled. And also the tongue, whiche is raysons exposytour, is deprived of his offyce, as it appereth in them, whyche are drunke, and them, whyche have grevous peynes in theyr head, procedynge of replecion. (52)

The 'fumosytie' that Elyot speaks of here echoes an earlier French text, Arnoul de Villeneuve's *Le Régime tresutile et tresproufitable pour conserver et garder la santé du corps humain* (*The Most Profitable and Useful Regimen for Maintaining and Guarding the Health of the Human Body*), published in Lyon in 1491, which is a commentary and explication of the *Regimen sanitatis salernitanum*, the famous Latin poem containing the caveats of the medical school at Salerno. Like Elyot, Villeneuve paraphrases Galen in his commentary on the negative effects of undigested meat: '... des viandes indigerees et grosses en l'estomac sont elevees fumees grosses faisant empeschement au cerveau, car c'est chose necessaire, de ces matieres grosses font fumees dissoluees et eslevees' ('... from large, undigested meat in the stomach, powerful gasses rise up, hindering the functions of the brain, since these materials necessarily produce dissolved and rising gasses') (6–7). Similarly, in his 1607 translation of the *Regimen sanitatis salernitanum*, the Elizabethan courtier Sir John Harington wrote: 'Great harmes have growne, & maladies exceeding, / By keeping in a little blast of wind: / So *Cramps & Dropsies, Collickes* have their breeding, / and *Mazed Braines* for want of vent behind.' (79; *Quatuor ex vento veniunt in ventre retento,/ Spasmus, hydrops, colica, vertigo, quatuor ista* [160]).

From these texts, which are explicitly based on the works of Galen and Hippocrates, it is apparent that, from the point of view of fifteenth- and sixteenth-century medicine, the body was not composed of discrete systems that were independent of one another. Far from belonging to a Bakhtinian 'lower material bodily stratum' that was cut off from the activities of the brain and the mind, the stomach and the bowels could become dangerous repositories of vital humors that fueled and deranged the brain, if the individual were dedicated to a life of dietary excess, like the young Gargantua. Thus Rabelais's description of the giant's early education is not merely pedagogical in nature, nor is it simply an allegorical comment on the necessary purgation of a brain that has been crazed by half-baked theology and religion. It is also a quite literal exposition of the medical theories of the ancient world as they were revised and revisited by Humanist scholars. Thus while the allegorical dimension of Rabelais's text often takes precedence over its literal level, there are also theoretical texts, such as the medical treatises cited here, that provide simple interpretations for seemingly scandalous and outrageous passages. Despite their apparent simplicity, however, these more concrete explanations serve to reinforce and to underline what are seemingly the most remote figurative interpretations.

With this medical conception of the body as a 'holistic' entity in mind, in which the humoral interplay between the stomach and the brain determines, to a certain extent, the spiritual well-being of the individual, one may explicate another passage in Rabelais in which repletion and constipation give rise to narrative excess and improvisation, which may nevertheless be read from an allegorical point of view that highlights Rabelais's Humanist, and to a certain degree Evangelical, sympathies. The penultimate chapter of *Pantagruel* details the 'excremental crisis' (Stephens 276), that the title character suffers after the conquest of the Dipsodes, and after the episode in which Alcofrybas spends six months in the giant's mouth. Undoubtedly these two passages have to be considered together as a logical progression, or perhaps even as a 'causal' sequence of events. The adventure/travel narrative of Alcofrybas's trip among the mountains and valleys of Pantagruel's teeth and throat logically leads to the proto-science fiction descent into the giant's belly, complete with pseudo-realist descriptions of the copper diving bells (or 'apples,' in Rabelais's terms) used for the expedition, the lamps that served to illuminate the giant's gullet, and the 'immunization' against noxious vapors used by the divers. While Rabelais's description parodies passages from Lucian and Galen (Antonioli 154–56) it also looks forward to ground-breaking stories by Poe and H.G. Wells.

As in the episode of Gargantua's education, the diptych that explores Pantagruel's anatomy from high to low is essentially concerned with the notion of *healing* an ailment caused by the 'humoral repletion' that is the consequence of eating too much meat. On its most basic and literal level, therefore, these chapters are a kind of hygienic improvisation and commentary, in the ancient sense of the medical discipline that understands the causes of disease as alimentary excesses and mistakes. But beyond this literal level, chapters 32–3 weave together an astonishing number of allegorical narrative strands that have been examined at length by Rabelais scholars. As in the case of the other passage examined here, Rabelais's text poses the question as to the nature and meaning of literal, physical purgation in the context of Evangelical Humanism. Is purgation always a metaphor for spiritual purification, or may we read these passages without undertaking the seemingly necessary allegoresis of the kind that the best readers of Rabelais have revealed to us?

The first element that must be taken into consideration is the placement of these two chapters within the general 'design' of *Pantagruel*. Considering the theological bases of even the coarsest descriptions in this work, which have been examined so convincingly by Duval, Persels, and others, it is clear that Pantagruel's illness originates in, or at least comments upon, the doctrinal impurities that upset the 'digestion' of the true faith. In order to elucidate this meaning we must isolate what implicitly seems to be the cause of Pantagruel's ailment. The text gives us two possible clues as to its origin, the first being a sylleptic play on an ancient text, borrowed from Budé:

> Quand [les médecins] furent en l'estomach, chascun desfit son ressort et sortirent de leurs cabanes, et premier celluy qui portoit la lanterne, et ainsi cheurent plus de demye lieue en un goulphre horrible, puant, et infect plus que Mephitis, ny la palus Camarine, ny le punays lac de Sorbone, duquel escript Strabo. (335)[5]

> When they [the doctors] were inside the stomach [Quand furent en l'estomach], each one released his spring and they came out of their cabins, and first the man bearing the lantern, and thus they fell over half a league into a horrible gulf, more stinking putrid than Mephitis, or the marsh of Camarina, or the foul lake Sorbonne, which Strabo writes about. (243)

The Scholastic theology of the Sorbonne is explicitly compared here to the stench of a stagnant lake, which is the source of the plague in the kingdom of Pantagruel's mouth in the preceding chapter, but which is also a metaphor for the stench of the garlicky 'ichors' rising from the pit of the giant's sickly stomach and bowels. The ills of the Utopian realm within the giant's mouth are thus caused by the 'bad air' that rises from this doctrinal lake, which is also literally the foul repository of the humors that have dangerously accumulated in the giant's stomach, if we diagnose his illness from a Galenic perspective. Moreover, it is the repletion of his body with this bogus doctrine that causes the giant to be sick, at least allegorically speaking.

Similarly, for Duval, these chapters represent a set of variations on the standard corporeal metaphors of the *corpus reipublicae mysticum* and the *corpus Ecclesiae mysticum*:

> In the context of an epic design that we have found to be both Evangelical and political, representations of the giant Pantagruel's body... inevitably recall the most familiar, canonical metaphors for political and ecclesiological organization – those of the state as the 'body politic' of the king, and of the Church as the

mystical 'body of Christ.' (125)

In an impressive display of erudition, Duval concludes that these final chapters represent the virtual apotheosis of Pantagruel as the Christ figure whose coming is the culmination of the work as an Evangelical, Humanist epic. Nevertheless, Pantagruel's illness intimates the existence of a flaw in the body politic, leading the reader to surmise that the ideal world toward which the work tends 'requires human correction' (134), meaning that the text ends with a call to Rabelais's contemporaries to work so that France may one day attain the ideal that is embodied in the body politic's head, the Humanist prince, who is also an embodiment of Christ. It would be difficult to dispute such a carefully argued interpretation of *Pantagruel*'s final chapters. Nevertheless, they contain certain elements that may be further elucidated from the point of view of the medical discourses discussed here.

It could be that the ultimate interpretation of these two puzzling chapters is related to the medical theories of the relation between the stomach and the seat of consciousness in the brain, which achieves its ultimate expression through the orifices of the mouth and throat, which Antonioli has called the 'carrefour entre le petit Monde de l'Homme et le Grand Monde, voie sacrée du souffle et de l'aliment, porte de la vie et voie d'accès vers les mystères du Monde intérieur' ('crossroads between the little World of Man and the Great World, the sacred way of breath and of sustenance, the door of life and point of access to the mysteries of the interior World') (155). If Pantagruel's mouth, and by extension his head, are indeed symbols of the monarch, and hence of France itself, as both Duval (130–31) and Defaux (412) demonstrate, and if his stomach is a diseased member of the French body politic that requires action of its subjects, then the supposedly 'normal' and 'banal' state of affairs in Pantagruel's mouth – a post-lapsarian world in which laborers subsist by the sweat of their brows, brigands beyond the mountains attack travelers, the population is constantly threatened by the plague, and sophistic charlatans such as Alcofrybas get paid for sleeping – is evidence of the 'mazed brain' of the country's ruling class, caused by the carnal repletion and constipation of the flawed religious doctrine issuing from the swamp of the Sorbonne, but also from the erroneous politics practiced by Pantagruel after the defeat of Anarche. As Duval remarks, the meaning of these metaphorical passages is both religious and political, as is the source of the giant's illness. The literal cause of Pantagruel's excremental distress is the amount of meat and garlic that he has eaten at Anarche's wedding feast, organized by the Devilish Panurge. Alcofrybas describes the cause of the plague in Pantagruel's mouth as follows:

> Lors je pense et calcule, et trouve que c'estoit une puante halaine qui estoit venue de l'estomach de Pantagruel alors qu'il mangea tant d'aillade, comme nous avons dict dessus. (332)

> Then I thought and calculated, and decided it was a stinking breath that had come from Pantagruel's stomach when he ate all that garlic sauce, as we have said above. (240)

The 'above' refers to Anarche's wedding banquet in chapter 31:

> Deux jours aprés, Panurge le maria avecques une vieille lanterniere, et luy mesmes fist les nopces à belles testes de mouton, bonnes hastilles à la moustarde, et beaulx trivars aux ailz, dont il envoya cinq sommades à Pantagruel, lesquelles il mangea toutes tant il les trouva appetissantes, et à boire belle piscantine et beau cormé. (329)
>
> Two days later, Panurge married him off to an old whore, and himself gave the wedding feast with fine sheep's heads, good slices of roast pork with mustard, and other good roast meat with garlic – of which he sent five loads to Pantagruel, all of which he ate, he found them so appetizing; and to drink, fine small wine and good sorb apple wine. (237–38)

As we have seen repeatedly, carnal excess is a symptom and metaphor of some form of religious, or in this case political, error. If, as Defaux has argued, Alcofrybas and Panurge are doubles of one another, serving as Rabelais's *sophista loquitur*, and immersing Rabelais's world in the labyrinth of sophistry,[6] then the ambivalence that appears in Alcofrybas's account of his travels in the giant's mouth is perhaps a parallel to the offhand and scandalous manner in which the triumphant, supposedly Humanist Prince treats his defeated foe in chapter 31, at the instigation of Panurge. Alcofrybas's description of his travels around the giant's mouth (and hence his brain?) is the linguistic double of Panurge's merry pranks, as the latter pumps the giant's stomach full of meat, causing a general disequilibrium in his body and mind.

In other words, the entire scene that takes place in Pantagruel's mouth exhibits a kind of sophistry that leads the reader to believe that the towns of Larynx and Pharynx belong to a Utopian world, when in fact this world is full of the same suffering, depravity, and misery as the 'normal' world of current-day France, despite the 'bonne chère' that Alcofrybas claims to enjoy throughout the kingdom. This is not surprising since, as a narrator, Rabelais's anagram is notoriously unreliable, hiding his deceptions behind a parodic and obsessive concern for details and numbers (Defaux 415–16). The dystopia that reigns in Pantagruel's mazed mouth and brain, and the narrator's account of it, are caused by the political excess that mirrors the alimentary excess of the wedding feast. In a footnote, Demerson remarks on the stark contrast that distinguishes Pantagruel's victory and its consequences for Anarche from Gargantua's victory over Picrochole: 'Dans le *Gargantua*, les géants auront réfléchi avec plus de sérieux et plus d'humanité sur le traitement qu'on doit réserver aux prisonniers' ('In *Gargantua*, the giants will reflect more seriously and in a more humane manner on the treatment that one should reserve for prisoners') (342 n.8). In contrast to what Duval calls the 'absolute perfection' of Pantagruel as the head of the idealized City of God (134), in these chapters it seems, on the contrary, that the realm described in Pantagruel's mouth is absolutely corrupted by the excesses of a prince who exhibits the most blatant of anti-Humanist and anti-Christian tendencies, most noteworthy among them being the lack of charity toward a defeated enemy, and the symbolic gluttony that is its consequence. In the Utopia that is Humanism's promised land, defeated monarchs simply cannot be debased to the level of ambulant mustard salesmen who are beaten by their prostitute wives, as Anarche is here. It is ultimately the 'bad meat' of this hyperbolic and unacceptable image that Pantagruel is unable to digest and expel.

The second cause of the giant's illness is much more scatological in nature. At this pivotal moment of the text, Rabelais's anagrammatic personification acts as a passive observer in the dystopia of the giant's mouth, enjoying himself amidst the general misery of the population. The interplay between the 'author' and his creation is a fascinating one if interpreted allegorically. From a Humanist point of view, the mere description of the state's putrid condition results only in the production of more refuse, or of more 'doctrinal shit,' as Persels would have it. The ambivalent nature of both the author's persona and of his main character cannot be overlooked, and produces a multiple perspective that makes it nearly impossible to determine exactly what the meaning of these passages is. From being a passive observer of what happens in Pantagruel's mouth, Alcofrybas becomes a producer at the chapter's end, in that his narrative account of his journey engenders two phenomena that are privileged throughout Rabelais's work. The first of these is important in the context of Persels's reading. Alcofrybas deposits his fecal matter in Pantagruel's *mouth*, operating a series of significant inversions: the refuse of the lower body enters the upper body; the end product of the digestive process enters the beginning of the digestive tract; the 'wisdom' that should issue forth from the mouth of the good Christian is stuffed back in this orifice as filth and doctrinal drivel. Secondly, Alcofrybas's account makes Pantagruel laugh, with all of the ambivalence that this entails. The giant's subsequent illness, caused perhaps in part by what Alcofrybas has made him eat, is evidence enough that laughter is not the appropriate response to the dire situation of the inhabitants living in Pantagruel's mouth (which is also that of most of Rabelais's readers), trapped in the putrid orifice of the giant as monster. Similarly, Pantagruel is both devouring ogre and suffering victim in these two chapters, and he will ultimately require Rabelais to assume one of his other alter egos at the end of the sequence, that of the Humanist doctor, who is a specialist in bodily purgations and their concomitant cerebral purifications. The giant's purgation here is indeed spiritual, but the necessity of purging the body politic of its social ills is much more important in these chapters. What is often lost in these kinds of readings is that both of these allegorical meanings are communicated to us via the literal, if parodic, description of a medical procedure.

One could hardly end a discussion of the medicine of scatology without mentioning the importance of vomit in these medical procedures. Aldebrandin de Sienne's chapter on purgation is followed and complemented by a chapter on vomiting: 'Por coi il fait bon user le vomir' ('Why it is Good to Use Vomiting') (56). None of the readers of chapters 32–3 of *Pantagruel* whom I have consulted have remarked that the cure for Pantagruel's illness is accomplished by oral evacuation, as opposed to defecation:

> Aprés, en tactonnant et fleuretant, [les pionniers] aprocherent de la matiere fecale et des humeurs corrumpues. Finablement trouvèrent une montjoye d'ordure. Lors les pionniers frapperent sus pour la desrocher et les aultres avecques leurs pasles, en emplirent les corbeilles: et, quant tout fut bien nettoyé, chascun se retira en sa pomme.
>
> Ce faict, Pantagruel se parforce de rendre sa gorge, et facillement les mist dehors, et ne se monstroyent en sa gorge en plus q'un pet en la vostre, et là sortirent hors de leurs pillules joyeusement
>
> Il me souvenoit quand les Gregeoys sortirent du cheval en Troye. Et par ce moyen fut guery et reduict à sa premiere convalescence. (335)

> Afterward, groping and smelling their way, they approached the fecal matter and the corrupted humors; finally they found a great heap of ordure. Then the pioneers beat on it to break it up, and the others, with their shovels, filled the baskets with it; and when all was well cleaned out, each one got back into his globe. That done, Pantagruel makes an effort to throw up, and there they came joyously out of their pills – it reminded me of when the Greeks came out of the Trojan horses – and by this means he was cured and restored to his earlier health. (243)

The semiotic incompatibilities that are typical of comic discourse underscore Rabelais's epic pretensions, which have been discussed by Duval throughout his careful descriptions of the design of Rabelais's works. There is no doubt that Rabelais is manipulating the standard metaphors of the body politic and the body of the monarch as Christ figure in these chapters. Nevertheless, it is also clear that he is inscribing, and mocking to a certain extent, the common medical procedure of induced vomiting, described in detail by Paré, including the preparation of pills that must be swallowed and vomited by the patient. This practice is expressed here in a Homeric comparison that is hardly appropriate to the elevated tone of the Greek epic.

The healing of the body politic's diseased members is nonetheless a heroic task, which must be entrusted to enlightened Humanist physicians such as Dr Rabelais himself. In fact, the subtext of these final chapters is contained in false cosmological details that are characteristic of Rabelaisian discourse, and which develop a symbolic register of physical healing that serves as a parallel to spiritual healing. The first image is that of the hot springs that are born of Pantagruel's 'chaude pisse'. As the Prologue of Rabelais's contemporary and patron Marguerite de Navarre's tale collection, the *Heptaméron*, makes evident, hot springs were places of literal, physical healing that literary texts consistently used as metaphors for the spiritual healing that was associated with religious pilgrimages, such as the one undertaken by Oisille at the beginning of Marguerite de Navarre's text. In passing, one might note that images of pilgrimages and rivers of urine are repeated in chapter 38 of *Gargantua*, in which the giant eats six pilgrims in his salad, echoing chapter 32 of *Pantagruel*. Similarly, the globe that adorned the Church of the Holy Cross in Orléans became, in Rabelais's text, a reminder and symbol of the epic struggle to purge Pantagruel of the humors disturbing his stomach: 'Et de ces pillules d'arin en avez une à Orleans, sus le clochier de l'esglise de saincte Croix'

(335) ('And of those brass pills you have one at Orléans on the steeple of the Eglise de Sainte-Croix' [243]).

In conclusion, then, it is evident from the scholarship devoted to Rabelais that his texts ultimately depict the Evangelical preoccupation with salvation and the true faith, and the need for good Christians to work toward the ideal Utopia of Christ's kingdom. It is also evident, however, that his works interrogate and expose medical theories in the most basic and literal manner imaginable, and that Rabelais was not concerned with physical healing merely as a metaphor for the life of the spirit. As a good doctor, he was concerned with the health of the body as an end in itself as well. Moreover, *Pantagruel* contains a political commentary on the need for the Humanist Prince to be magnanimous toward his enemies and toward the body politic that looked to him for leadership. Thus the health of the body that was the essential goal of medicine was also a metaphor for the health of the state in Rabelais's text, as well as an allegory for the life of the spirit. It might be argued, however, as I have here, that one of the most important subtexts of Rabelais's work is the exploration of medical theories, and the literal link that they established between the bowels and the brain, the lowest bodily functions and the highest strivings of the human mind.

Notes

[1] All page references to Rabelais are from Mireille Huchon's edition. English translations of Rabelais are from the Donald M. Frame edition; all other translations are mine.

[2] 'La cueue est caude de se nature, et est pesans à l'estomac, et donne abhomination, et taut l'appetit de mengier, et li humeurs ki en est engenree est cole rouge' (The tail is hot by nature, and weighs on the stomach, producing abomination, and killing the appetite; the humor that it produces is red bile) (138).

[3] 'On l'a dit : la régulation de la santé par le régime et l'association de la diététique avec la médecine et le morale (qui elles-mêmes ont partie liée) sont des usages anciens, que les humanistes n'inventent pas. L'innovation, à la Renaissance, tient moins à la pratique qu'à la promotion et la diffusion du principe parmi les cercles lettrés, comme objet d'étude. La diététique devient une question à la mode, une discipline qui appartient désormais au bagage culturel de l'honnête homme (We've said this before: the regulation of health by means of diet, as well as the association of nutrition with medicine and morality are ancient practices, which the Humanists do not invent. The innovation of the Renaissance in this regard has less to do with these practices themselves than with the promotion and diffusion of these principles in learned circles, as objects of study. Nutrition becomes a fashionable topic, and a discipline that would belong from that moment onward to the cultural baggage of any honest gentleman) (81–2).

[4] Aside from various mixtures of herbs, one which also includes white hellebore, aloe, and myrrh, Paré speaks of pills containing scorpion oil, which induced strong vomiting, and a kind of 'brew' made of anchovies ('saumure

d'anchois') that a gentleman from Navarra recommended to him. In Paré's text, purgation is accomplished both through vomiting and defecation, while purgative drugs had to be specific to the individual: 'Et les faut diversifier selon qu'on connoistra l'humeur pechant: aussi en prenant indication du temperament du malade, de l'aage, coustume, pays, saison de l'année, sexe, air ambiens, et plusieurs autres choses semblables, qu'on verra estre necessaires, et principalement de la vertu' (And one must diversify these [drugs] according to the diagnosis of the offending humor, and by indicating the temperament, age, customs, country, and sex of the patient, as well as the season of the year, the ambient air, and other similar things, which one will judge to be necessary, especially as concerning virtue) (414).

[5] On this passage's relation to Budé, see Demerson's note: 'Strabon parle bien du lac Serbonis; Budé, dans une lettre à Érasme, reproduit la même étymologie facétieuse que Rabelais' (Strabon is referring to the Lake Serbonis; in a letter to Erasmus, Budé reproduces the same facetious etymology as Rabelais) (349 n.8).

[6] '...reconnaître la parenté de Panurge et d'Alcofrybas, c'est souligner du même coup l'importance du sophisme dans l'univers rabelaisien. Car Alcofrybas est le créateur de cet univers, il lui prête sa voix et son style, il le modèle à son image. Si chez Panurge la puissance de la parole est destinée à faire triompher une *libido dominandi* tyrannique, elle débouche chez Alcofrybas sur une création littéraire dont quatre bons siècles d'exégèses, de gloses et de commentaires contradictoires n'ont pas épuisé les ambiguïtés. (...to recognize the family relationship between Panurge and Alcofrybas is to underline by the same gesture the importance of sophism in Rabelais's universe. Since Alcofrybas is the creator of this universe, he lends it his voice and his style, and he models it according to his own image. If, for Panurge, the power of the word is destined to bring about the triumph of a tyrannical *libido dominandi*, this same power leads to Alcofrybas's literary creation, whose ambiguities four long centuries of exegesis, glosses, and contradictory commentaries have not exhausted) (416–17).

Chapter 4

'The Mass and the Fart are Sisters': Scatology and Calvinist Rhetoric against the Mass, 1560–63

Jeff Persels

It seems that in 1526, in the Dominican monastery at Auxerre, there was an old pox-ridden monk whose age and infirmities had long prevented him from celebrating mass. Determined nevertheless to honor his God on the Feast of Corpus Christi, he participated in the ceremony and partook of the consecrated bread and wine. Shortly thereafter, en route to his cell, he was suddenly taken ill and vomited in front of the chapter room door. When the holy mess was found, it provoked a hue and cry among the brethren, who were troubled as to how to proceed in the prior's absence. They first protected the sacred vomit from the footsteps of the inattentive passer-by and the appetites of the monastery dogs by means of lit candles and the tabernacle, normally reserved for saying masses for the dead, all the while singing hymns and anthems appropriate to the day. Custom and church teaching dictated that one from among them be so brave as to consume the regurgitated Host. No friar finding himself worthy of so great an honor, it was decided that the consecrated egesta should at least be reverently taken up and transported to the church, along with the soil on which it fell, accompanied by the brethren in full expiatory procession. Once inside the church, the elders concluded that it would be preferable to incinerate the hallowed remains together with the affected soil, and to preserve the ashes in a reliquary, rather than to let the body and blood of Our Lord putrefy and decompose any further.[1]

This anecdote, along with numerous and vociferous interventions on the part of its first-person narrator, Calvinist theologian Pierre Viret, was presented as footnoted commentary in a ferociously satirical 1554 'translation' of a 1520 Roman missal.[2] What is most noteworthy about it, at least for the purposes of the following argument, is that far from being dismissable as anomalous anti-Catholic propagandistic rant, it can be considered representative of a discursive vernacular style cultivated by contemporary Reform proselytizers, many of whom, including and especially Viret, were orators and polemicists of the first water.

Of late, historians of religion and literary critics have turned their attention to re-examining the form and mechanics of sixteenth-century religious polemic, understandably most abundant with regard to the seminal Latin and vernacular Lutheran tracts of the 1520s and 1530s.[3] Such attention has made significant effort to correct the somewhat awkward neglect of various Evangelicals' and Reformers' frequent recourse to gross example and crass expression, not just in heated, often *ad hominem* rebuttals but also, as in this instance, in the very vulgarization of difficult doctrinal issues considered vital to the salvation of the faithful and to the reform of contemporary religious institutions and practice. Rarely is such rhetoric simply abusive – a convenient and conveniently direct or graphic attempt to dismiss an adversary and his argument. Infrequently it is resorted to merely for shock value, a deliberate forced intrusion of low style or a calculated disruption of accepted discursive levels to oblige the reader or listener to sit up and take notice. Most often it goes beyond that and is deployed in earnest to enable the minister of God's word to fulfill his primary pastoral function, to 'make Scripture easier' ('rendre les Escritures plus faciles') as Viret himself says (*Dialogues* 490), and, as Francis Higman long ago pointed out most tellingly about Calvin's own polemical works, 'to convince rather than to prove' (*Style* 41).

Few things, it would seem, so convincingly and viscerally (taken both literally and figuratively) convey the primary and pressing concerns of the *religion prétendue réformée*, most effectively and succinctly the eucharistic and soteriological drama of the Reform crisis, than the rhetoric of elimination and purgation, oral and anal. If, as an anonymous 1561 Calvinist satire, the *Sentence decretale, et condemnatoire au fait de la paillarde Papauté* (*Decretal and Condemnatory Sentence Regarding the Debauched Papacy*), would have it, 'the mass and the fart are sisters' ('la messe et la vesse sont sœurs') (D4^{v}), then scatology and Reformist polemic, be it Evangelical, Lutheran or Calvinist, are brothers – brothers-in-arms, that is.

I would like here to use Viret's anecdote to begin to build a case for recognizing the integral nature of scatological rhetoric in French as part of the Calvinist proselytizing agenda, and to build a case against a perceived tendency to blush for it and go on: to gloss over it, to trivialize it by reducing it to a matter of personal style, or to dismiss it as an inconsequential and uncouth by-product of a less aesthetically sensitive or evolved age. For initial and authoritative inspiration, we need look no further than the 1541 French translation of so severe a theological classic as Calvin's *Institutio*. Though rather prim by Lutheran standards, and, no lover of the 'scandalous' Rabelais, Calvin, taking his cue from so august an authority as St Paul, does seem to appreciate the aptness and usefulness of eating and digestion as analogical figure. From the dedicatory epistle to Francis I:

> Why do [our Catholic adversaries] fight so fiercely and cruelly for the mass, purgatory, pilgrimages and like trifles? ... Why, I ask, unless it is *that they take their belly for their God, their kitchen for their religion*?[4] Take these away and they not only no longer think that they can be Christian but no longer believe they can be men. For though some feast in splendor and others merely survive by gnawing a crust of bread, yet all eat out of the same pot, which without their vigilance would not only grow cold but would freeze over. *He among them who looks best after his own belly is considered the most zealous in faith.* In short, they

> all have the same goal, whether it be keeping their power or filling their bellies. (*Institution* 1.32–3, emphasis added)[5]
>
> Pourquoy combattent-ils d'une telle rigueur et rudesse pour la Messe, le Purgatoire, les pélerinages et tels fatras? .... Pourquoy, dy-ie sinon pourtant que *leur ventre leur est pour dieu, la cuisine pour religion*? Lesquels ostez, non seulement ils ne pensent pas qu'ils puissent estre Chrestiens, mais ne pensent plus estre hommes. Car combien que les uns se traitent délicatement en abondance, les autres vivotent en rongeant des croustes, toutesfois, ils vivent tous d'un pot, lequel sans telles aides non seulement se refroidiroit, mais géleroit du tout. Pourtant *celuy d'eux qui se soucie le plus de son ventre, est le meilleur zélateur de leur foy*. Bref, ils ont tous un mesme propos, ou de conserver leur règne ou leur ventre plein.

From 'Des sacrements'('On the Sacraments'):

> But we will continue more appropriately in another chapter our comparison between circumcision and baptism. The ablutions and purifications showed them *their unspeakable squalor, their filth, their pollution*, by which they were in their very nature *sullied, stained, and loathsome*, yet also promised them another *ablution* by which they would be *purified and washed clean of their stains and defilement*. And this ablution was Jesus Christ, by whose blood we are *purged and made clean* [I John 1.7 ; Apoc. 1.5], by whose wounds we are healed [I Peter 2.23], such that our filth is covered in order that we may bear a true purity before God. (*Institution* 4.311, emphasis added)[6]
>
> Mais nous poursuyvrons en un autre passage plus proprement la comparaison de la Circoncision et du Baptesme. Les ablutions et purifications leur démonstroyent *leur immondicité, leur ordure, leur pollution*, par laquelle ils estoyent en leur nature *souillez, maculez, et infects*, et aussi leur promettoyent un autre *lavement*, par lequel ils seroyent *purifiez et nettoyez de leurs macules et infections*. Et ce lavement estoit Iesus Christ, par le sang duquel nous sommes *purgez et mondifiez* (I Jean 1, 7 ; Apoca. 1, 5), par les playes duquel nous sommes guériz (I Pier.2, 23), tellement que noz souilleurs sont cachées, afin que nous apportions une vraye pureté devant Dieu.

From *Contre les Anabaptistes* (*Against the Anabaptists*), 1544:

> In the end, just like a drunkard who, after a good belch, throws up the vile broth that weighs down his stomach, so these wretches [i.e., magistrates] who, after slandering this holy office that our Lord so honored, finish by vomiting up much worse blasphemies. (*Opera* 7.90–91; quoted in Higman, *Style* 148)
>
> En la fin, comme un yvrongne, apres avoir bien rotté, desgorge le villain brouet qui luy charge l'estomac: ainsi ces malheureux, après avoir detracté de ce sainct estat que nostre Seigneur a tant honnoré [i.e., magistrates], vomissent finalement à pleine gorge des blasphemes beaucoup plus desordonnés.

If such implied license to characterize the fallen church's current condition and its fallen flock's eternal one as a roll in the gut or the gutter should prove insufficient, Calvin provides explicit rhetorical instruction elsewhere. From *Des scandales* (*Of Scandals*), 1550:

> ... it pains pretentious people given to ostentation that the Holy Spirit uses a *vulgar and simple language.* Those educated in the liberal arts and who are accustomed to a pure and elegant style reject or scorn this way of speaking as being *too crude and unpolished*, ... *Jeremiah's style smells of the barnyard, and Amos' of the sheepfold*; I freely confess that the books of our religion, in which is enclosed the heavenly wisdom of God, not only are not made-up with rhetorical colors but do not at all have the appearance that learned people might prefer. (64–5, emphasis added)[7]

> ... il fait mal à gens enflez et addonnez à ostentation que le Sainct Esprit use en l'Escriture saincte *d'un langage grossier et simple.* Ceulx aussi qui ont esté instruicts aux sciences humaines et sont accoustumez à un style pure et elegant rejettent ou mesprisent ceste façon de parler, comme *trop rude* et *mal polie*,.... *le style de Jeremie sent son païsant, celuy d'Amos son bergier*, je confesse volentiers que les livres de nostre religion, où est comprinse la sagesse celeste de Dieu, non seulement ne sont point fardez de couleurs rhetoriques, mais aussi qu'ils n'ont point une telle apparence que peuvent desirer gens lettrez.

If the original style of the sacred truths themselves smells of the barn and the sheepfold, what better justification for retransmitting the revivified truth in equally 'vulgar' and 'simple' French translation?

Such potentially malodorous rhetoric – malodorous, in reality, because lacking in rhetorical adornment – was perceived, it could be argued, as part and parcel of the ministry's sacred obligation to the faithful, and early Calvinist reformers, as we know well, took the role of the ministry very seriously: 'The pastor must follow the faithful closely, must share their life so that he may know it intimately; and he must strive to direct each of them according to Gospel teachings' ('Il faut que le pasteur suive de près les fidèles, qu'il se mêle à leur vie, qu'il en connaisse les circonstances et qu'il s'efforce de diriger chacun d'entre eux selon les préceptes de l'Evangile') (Barnaud 494). Just as their prophetic, apostolic and patristic forebears of the 'primitive Church' to whom they most often appealed, they recognized and exploited early on, and much to the Roman Church's dismay, the efficacy of preaching to the barnyard in barnyard terms.[8] Few did this better than Pierre Viret, in tract after tract, not, however, because his work is lacking in rhetorical adornment, but because he, like Calvin, apparently believed that he was exercising that most Christian and Pauline of paradoxes, putting the low to a high purpose, turning scatology into rhetoric, excrement into eloquence, reminding his audience of the role of vomit and its potential riches both in the *vox christi* and the *vox populi*.

Before taking up the specifics of Viret's anecdote, it might be useful to recall briefly the earthiness of popular faith and its expression, regardless of confessional persuasion. Such anecdotes as Viret's, however adapted to specific polemical use, were far from exceptional, even among writers with a less pronounced if nonetheless detectable political or religious agenda. Viret's and others' proselytizing efforts would have been necessarily attuned to this faith and its expression if they hoped to succeed on a mass scale. Representative examples may readily be drawn from among the orthodox *faits divers* of 1515 considered worthy of record by the anonymous chronicler(s) of the *Journal d'un bourgeois de Paris*. The first even bears a striking similarity to the Auxerrois incident:

> In the same year, it happened in the city of Poitiers that a madman dripped the precious blood of Our Lord onto the corporal of the altar while a priest was saying mass. The incident occurred after the consecration, as the priest was raising the chalice. The consecrated wine was white. Whereupon a beautiful miracle occurred. For no sooner had the wine fallen on the sacred corporal than it turned red. It was immediately, with great diligence and reverence, taken up by the priests, and a reliquary was made for it so that it might be worshipped. The criminal was condemned to be locked away rather than executed, for reason of insanity, and it was ordered that he never be set free and have only bread and water to sustain him. (35)
>
> Audict an, il advint en la ville de Poictiers que un homme fol fit cheoir sur les corporaux estant sur l'autel le precieux sang de Nostre Seigneur, lorsqu'un prestre chantoit la messe, et fut ce faict après la consecration comme ledict prestre levoit le calice; et estoit la consecration de vin blanc; dont il advint un beau miracle, car sitost qu'il fut tombé sur les corporaux sacrez il devint rouge; et incontinent fut recueilly en grande diligence et reverence par prestres. Et en fut faict reliquaire pour adorer; et fut le malefacteur condamné à estre encloz entre deux murs sans le faire mourir, à cause qu'il estoit insencé, et fut dit que jamais n'en partiroit et mangeroit pain et buroit eaue seullement.

And from 1522, also in the Poitou:

> … a great and marvelous event occurred in the region of Poitou. It happened that a band of adventurers was going to war in the Guyenne, three of whom wanted to rob a local church. When they entered, they found a priest and asked him to go up to the altar and give them the holy ciborium or, at least, the chalice. He refused, saying gently that that was where God lay and that he would not take his resting place from Him. Thereupon the adventurers furiously slew the priest, running a sword through his body, climbed up to the altar and took the holy ciborium and the Host, which all three of them ate, and also carried off the chalice. They then went to the tavern, where they saw their companions and confessed what they had done. The criminals were unable either to eat or drink, saying that they were drunk on what they had eaten in church. Then their companions noticed that they began to smoke and that great fumes were coming out from every part of their bodies, such that it seemed they were burning in a furnace. The companions became so frightened that they fled the tavern, leaving the criminals there, and a short time later, when they came back to check on them, they found them completely burned. And no one knows how they burned, unless it had been the will of God, as punishment for the evil that they had done. (138–39)

> ... il estoit advenu au païs de Poictou une grande et merveilleuse chose. C'estoit qu'il y eut une bande d'advanturiers qui alloient à la guerre en Guyenne, desquelz y en eut troys qui vouleurent aller desrobber une eglise parrochialle de là; et eulx, entrans dedans, trouverent un prestre auquel ilz dirent irreveramment qu'il montast là hault, pour leur bailler le sainct ciboire, au moins la couppe, ce qu'il refusa en leur disant doulcement que c'estoit où Dieu reposoit et qu'il ne l'osteroit point. Lors incontinent furieusement ilz l'occirent et luy traverserent une espée au corps et monterent et prindrent le sainct ciboire et le *Corpus Domini*, le mengerent tous troys et emporterent ladicte couppe, puis s'en allerent à la taverne vers leurs aultres compaignons, leur declarant ce qu'ils avoient faict; et iceux ne peurent boire ne menger, et disoient qu'ils estoient tous soulz de ce qu'ilz avoient mengé à l'eglise. Lors leur bande apperceut qu'ilz commencerent à fumer et que une grande fumée sailloit de leur corps de tous costez, tellement qu'il sembloit qu'ilz fussent en une fournaise; dont les aultres en eurent si grand peur qu'ilz s'enfuyrent et les laisserent là, et peu après revindrent pour les voir, mais ilz les trouverent tout bruslez; et ne sçait-on comment ilz furent bruslez, sinon que c'estoit le vouloir de Dieu, par punition du mal qu'ilz avoient faict.

Historians have argued convincingly from direct contemporary complaints that late medieval popular devotion was characterized by 'a belief in the objective presence of the divine in material elements' (Eire 17), provoking charges of idolatry and resulting in acts of iconoclasm on the part of Evangelicals and Reformers.[9] The above passages drawn from the *Journal* not only bear irrefutable witness to that belief and its consequences but, unlike Viret's, do so with something approaching what we might call journalistic objectivity, at least as close as good factional chronicles could come in a partisan age. Both examples focus, conveniently enough for reading in the context of Viret's anecdote, on the 'material elements' of the Eucharist, the wine and the bread, though the 'objective presence of the divine' was scarcely limited to these most central and defining ones. It is to this belief that Viret as exemplary Calvinist polemicist addresses his appeal; the subtlety of his rhetorical transformation of that presence into excrement, that most material of elements, as we shall see, can only be fully appreciated in light of it. But it is also against this belief that Viret, as exemplary Calvinist polemicist, does scatological battle. The external rites that he and many of his contemporaries criticized in Catholic practice are internalized, so to speak, by recourse to an instantly accessible and literally visceral trope: they pass into the body, are digested (incompletely or not a all) and returned mortal, unclean, unholy, unmasked for what the Reformers thought them to be – so many empty symbols with no real link to the immanence they claimed.[10]

Viret's anecdote comments, of course, on the issue at the very heart – or, as Viret would have it, perhaps, in the very bowels – of the Catholic / Calvinist split, the literal and figurative readings of the Last Supper, the doctrine of the True Presence, the ultimate fly in the ointment or the hair in the soup at the 1561 Colloque de Poissy that would doom any French effort at reconciliation, let alone co-habitation:

> Of all the disagreements there are today among Christians in the matter of religion, there is not one that so troubles and devised them and that slows and

prevents the course of the Gospel and the union and harmony that should be among them, than that one which concerns the Last Supper of Our Lord and the mass of the Roman Church. (Viret, *Principaux poincts* 3)

Entre tous les differens qui sont auiourd'huy entre les Chrestiens en la matiere de la religion, il n'y en a point qui plus les entretienne en trouble & dissension, & qui plus retarde & empesche le cours de l'Evangile, & l'union & concorde qui devroit estre entr'eux, que celuy qui concerne la saincte Cene de nostre Seigneur, & la Messe de l'Eglise Romaine.

As one might expect, his own dismissal of transubstantiation is constantly reiterated, often with calculated use of current Scholastic idiom, such as in the following from *De la vertu & usage du ministere de la parolle de Dieu*:

> [The Church fathers] did not however mean nor wish to imply that there was a conversion of the bread and the wine into these [i.e., the body and blood of Christ], or a natural and substantial union between them, nor a connection made by corporeal and fleshly means, but rather a true spiritual communication, such as is appropriate to the nature of the Sacraments and to the spirit of God from which they draw their efficacy. (C4[v])
>
> [The Church fathers] n'ont pas pourtant entendu ny voulu donner à entendre, qu'il y eust une conversion du pain & du vin en iceux [i.e., the body and blood of Christ], ou une union naturelle & substantielle entre eux, & une communication qui sufist par maniere corporelle & charnelle, mais spirituelle, & veritable neantmoins, & telle qu'elle convient à la nature des Sacremens & de l'esprit de Dieu, duquel ilz prennent leur vertu.

If, for Viret and other reformers, the Catholic error leading into sin was essentially a rhetorical one, the literal reading ('by corporeal and fleshly means') of Christ's words at the Last Supper, of what the patristic writers had recognized as figurative ('but rather a true spiritual communication'), that error was compounded across the centuries by its being re-enacted *ad infinitum*, or, in this particular instance, *ad nauseam*.[11]

> This explanation [of Christ's words at the Last Supper] cannot work unless they be constrained to accept and to recognize rhetorical figures, at least that the grammarians call Synecdoche [Viret uses the term 'metonymie' elsewhere], that is, comprehension [Viret draws here on Latin etymology: 'to comprise, to include'], which means a manner of speaking by which we often mean and take the whole for the part, or the part for the whole; or the content for the form, or the form for the content; or the cause for the effect, or the effect for the cause, understanding both by either name; or either by both names. (*De la vertu & usage* 280)
>
> Ceste exposition [of Christ's words at the Last Supper] ne peut avoir lieu, qu'ilz ne soyent contreinctz d'y recevoir & recognoistre quelque figure, pour le moins, celle que les grammariens appellent Synedoche [Viret uses the term '*metonymie*'

> elsewhere], c'est à dire, comprehension, qui signifie une maniere de parler, par laquelle nous signifions & prenons souventesfois, ou le tout pour la partie, ou la partie pour le tout : ou le contenu pour ce qui contient : ou ce qui contient pour le contenu : ou la cause pour l'effect, ou l'effect pour la cause, comprenans tous les deux, par le nom de l'un ou de l'autre: ou l'un tant seulement, par celuy qui signifie tous les deux.

What more accessible way to point up what Calvinists believed to be the fatal absurdity of this most fatal of rhetorical slips than Viret's corporeal and carnal anecdote concerning Dominican vomit, provoked by ingestion of a bit of bread which, according to the doctrine of transubstantiation, is no longer merely bread and a drop of wine which is no longer merely wine, into a body steeped in sin?

On its first Genevan printing by Jean Girard in 1554, the *Cauteles et canon de la messe* constituted the last 300 pages of a more ambitious work retracing and challenging what Viret considered to be the 'misguided' evolution of the Catholic mass, from simple sacrament to a complex and 'idolatrous' ritual, now as foreign to the practice of the primitive church as Christ was distant from the communion wafer, in Théodore de Bèze's celebrated Poissy formulation.[12] The *Cauteles* reproduce a 1520 Latin text (Lyons: Jean de Cambray) of the Roman Missal, the very document to which apostate Italian monk Agostino Mainardo had already colorfully referred to by this date as 'a stinking and foul latrine in which the mass was finally locked up and where those who have business with it go to use it' ('un retrait puant & infect, auquel la Messe a este finalement enclose, & serrée, & ou ceux qui ont affaire d'elle, s'adressent pour s'en servir') (*vv^{v}). Viret sharpens his satirical axe on we might today call the 'Frequently Asked Questions' concerning the actual physical preparation and administration of the sacrament of the altar: 'things,' in Viret's words, 'which are to be observed by him who wishes to say mass and the dangers that can befall him' ('des choses qui sont à observer, à celuy qui veut dire Messe, & des dangiers qui y peuvent survenir') (568). The italicized Latin runs along the inside gutters, the remaining bulk of the page being taken up with what Viret assures us is a 'faithful' French vernacular translation, 'so that anyone may know better the great mockeries and abominations contained in the mass and in those who say it and attend it' ('affin qu'un chacun puisse mieux cognoistre les grandes moqueries & les grandes abominations qui sont en la Messe, & en ceux qui la disent & qui y assistant') (568). Should these mockeries and abominations not prove to be self-evident, Viret provides in addition 'some explanations and annotations, the better to clarify the contents' ('quelques expositions & annotations, pour mieux esclaircir les matieres') (568), in the form of copious chapter endnotes, carefully numbered, as Viret himself informs us, for our convenience.

This cumbersome but nonetheless efficient layout (for the reader if not for the printer) recalls, as it is intended to, that of traditional exegetical glossing. Though the Latin is present, it is typographically already less accessible than the accompanying French translation, most likely alerting us to an intended readership whose grasp of the vernacular is stronger than its knowledge of Latin but who can yet delight in the ridicule and discredit resulting in the move from the distant language of the Church to the familiarity of Viret's French. This move seems calculated to produce a specific effect, as much use of the vernacular was in Calvinist polemics. It was not just intended to render Scripture and doctrine more accessible, as do Pierre Robert Olivétan's Bible translation, Clément Marot's and Théodore de Bèze's renderings of the Psalms, and the French translation of Calvin's *Institutio*, and as Viret himself notes of his own work in the common speech:

> If it does not profit the learned, I hope that it will at least profit in some way the simple folk of our tongue, who do not understand Latin or who cannot follow the elevated style and the profound thought of those who have written on these matters, either more learnedly or more subtly and obscurely, or more briefly and succinctly than was required by the simple folk. (*De la vertu et usage* b3^{v})

> Si elle ne profite aux savans, j'espere qu'elle pourra profiter de quelque chose, pour le moins aux simples gens de nostre langue, qui n'entendent pas la langue latine, ou qui ne peuvent pas bien comprendre le haut stile, & la profonde doctrine, de ceux qui ont escrit de ces matieres, ou plus doctement ou plus subtilement & obscurement, ou plus brievement & succinctement qu'il n'estoit requis pour eux.

The shift to the vernacular, as most visually apparent here, was also arguably intended to desanctify the hallowed practices of the Church, much as we might experience today in attending supertitled opera, the English translation dulling the musical force by reminding us how inane and repetitive most libretti are. By rendering the 'Frequently Asked Questions' concerning the physical celebration of the mass into everyday French, Viret's intent is to sow the seeds of doubt through ridicule, onto which he then piles the fertilizer in his annotations. These make plain the irony of referring to the 'elevated style' and 'profound thought' of Catholic dogma, when what they result in is rather the low style and shallow doctrine illustrated by the Auxerrois Dominicans in what amounts to a scatological farce.

The anecdote is a mere footnote, number 3 to Chapter 21, 'Concerning uncertain and vomited gods of the Mass' ('Des dieux de la Messe douteux & vomis'), which contains two items only, brief enough to be cited in full:

> Also, if the host our a part of the host is found under the pall or under the corporal, and if it is not sure whether or not it has been consecrated, the priest must take it reverently, after having partaken of the blood, as you will find under the heading, 'Concerning the Celebration of the Mass.' Also, if the body of the Lord given to an invalid is thrown up ['rejected'] by him because of his illness or some other cause, let the priest receive it as diligently as possible.
>
> If through human weakness, there is no one willing or able to take what the invalid has thrown up, then let it be burned and its ashes preserved in a reliquary. (645–7)[13]

> Item, si l'hostie, ou une partie de l'hostie est trouvée sous le volet ou sous le

> corporalier, & si on doute si elle est consacrée ou non, le prestre la doit prendre reveremment, apres qu'il a pris le sang: comme tu le trouveras plus au long au tiltre, De la celebration de la Messe. Item, si le corps du Seigneur, baillé au malade, est par luy reiecté par infirmité, ou quelconque autre cause: qu'on le reçoive [Latin *recipiatur*] le plus diligemment que lon pourra.
>
> Et si par humaine fragilité, on ne trouve personne, qui vueille, ou puisse prendre [the Latin is *sumere*, to take, to lay hold of] ce qui aura esté vomy du malade [the Latin reads simply *rejectionem*, a throwing up], que cela soit bruslé, & les cendres soyent gardées au reliquaire.

Viret's artfully timed and placed anecdote teases out *ad absurdum* the implications of the spare Latin text and more suggestive translation. Reform polemicists, Viret among them, were, as we know, particularly good at tapping the weak points of an argument taken to its extreme, as often as not postulating that extreme themselves, in the face of actual orthodox belief. Factual, finessed or made up out of whole cloth – its customized fit is somewhat suspicious, however accurately it may mimic such 'factual' reports as that of the *Journal d'un bourgeois de Paris* cited earlier – this colorful *exemplum* bears the hallmarks of the Calvinist polemical model, as even a cursory examination of details can demonstrate, especially when read against those of the *Journal*.

Time: the very feast of Corpus Christi, of Christ's body, but from which Viret distances himself ('that *they* call Of the body of God ('qu'*ils* appellent Du corps de Dieu,' emphasis added), a papally conceived and instituted feast day, by Urban IV's 1264 bull 'Transiturus,' to celebrate the Host, the Blessed Sacrament of the Altar, pointedly confirmed by the Council of Trent in 1551.[14] Viret could not have chosen (nor been handed, assuming any truth to the tale), a more propitious day for taking on the doctrine of transubstantiation.

Setting: a Dominican monastery, the twelfth-century preaching and mendicant order, home fraternity of Albertus Magnus and Thomas Aquinas, the latter the author, of course, of the well-known hymn to the Blessed Sacrament for the Office of Corpus Christi, the *Pange lingua gloriosi*, included in the Roman Breviary (and worth a hundred days' indulgence to priests). Its penultimate stanza, the *Tantum Ergo*, and accompanying anthem provide Viret with material for ironic double-entendre.

Characters: the butt of anti-clerical humor since time immemorial, a monk whose corporeal decrepitude and spiritual ineptitude is a result of earlier venereal indulgence, hardly a fitting receptacle, reliquary or even conduit for the hallowed elements, or, conversely, for Viret's purposes, only too fitting a receptacle, reliquary, conduit for the Catholic 'dieu de paste,' or 'pastry god.' It is perhaps worth noting the contrasting behavior of both the quietly pious victim of the 1522 Poitou cautionary tale and the 1515 officiating priest in the *Journal* who, unlike Viret's protagonist, is *not* guilty of spilling (or spewing) the consecrated wine. The ignorant and idle multitude of brethren whose dithering lends an air of comic crisis and whose naïve obliviousness to the irony of the situation and the double meaning of the feast day's traditional office is mercilessly exploited, allowing Viret to enter into a generous and persuasive complicity with his audience.

> Then the novices were ordered to sing all day long the following verse of the hymn for this feast day, which begins: *Tantum ergo sacramentum, veneremur cernui* etc. That is, 'Let us honor, heads bowed, this great sacrament,' that is, this vomit from monsieur the beautiful father, which was such a noble and excellent sacrament. They also sang the day's anthem, which begins, *O sacrum convivium*, etc., that is, 'O the sacred banquet.' And not without reason. (52)

> Puis fut commandé aux novices de chanter tout le iour ce verset de l'hymne de celle feste, qui se commence, *Tantum ergo sacramentum, veneremur cernui &c.* C'est à dire, Honnorons donc estants enclinés & courbés en terre, ce sacrement si grand : c'estoit a dire pour lors, ce vomissement de monsieur le beau pere, qui estoit un sacrement fort noble & fort excellent. Ils chantoyent ausssi l'antien du mesme iour, qui se commence, *O sacrum convivium, &c.* c'est à dire, O le sacré banquet. Et non sans cause.

The Devil, so it is said, is in the details, and that seems precisely to be Viret's point. His translation and annotation of the Missal forces attention to such grossly material physical detail, the objectionable 'objective presence,' Devilish consequences of the Roman Church's refusal of a spiritual, figurative understanding, a corporeal entombment as opposed to the expected spiritual elevation of the Host promised by the very ceremony of the mass, particularly on this day of all days. All of this as if to say: 'How far we have come from the Lord's Supper! How far a literal, corporeal evolution of that concept has led us astray: from the *Hoc est corpus meum* of Christ to the *Hoc est rejectio mea* of a syphilitic Dominican friar! For a tradition that refused, almost without exception, the easy woodcut imagery of the Lutheran broadsheet, as Francis Higman has pointed out (*Piety and the People* 2–3), this is an astonishingly 'graphic' and graphically repugnant image: the sacramental, sacrificial, salutatory blood and wine reduced to a vile mess to be protected from the stray cur.

Viret's anecdote plays masterfully, as surely it was meant to, with the Gospel parable of digestion and its discontents: *non intellegitis quia omne quod in os intrat in ventrem vadit et in secessum emittitur quae autem procedunt de ore de corde exeunt et ea coinquinant hominem* ('Do you not see that whatever goes into the mouth passes into the stomach, and so passes on? But what comes out of the mouth proceeds from the heart, and this defiles a man') (Matt. 15.17–18). Not to mention, much more slyly, *sicut canis qui revertitur ad vomitum suum sic inprudens qui iterat stultitiam suam* ('Like a dog that returns to his vomit is a fool that repeats his folly') (Prov. 26:11; an analogy put to similar use by Calvin himself in the famous epistle to Francis I prefacing the *Institutes* [1.46–7]). Are not Dominicans the 'hounds of God'? His allusion to this bit of scripturally sanctioned scatology is so much shorthand for dismantling the doctrine of the Eucharist. The monk vomits up – and his brethren keep returning to – what he has taken in, the body of Christ. The doctrine of its transubstantiation is implicit cause of an upset stomach, on the annual Catholic feast day reserved for that body. That feast is a recent one, instituted in 1264, confirmed by Clement V in 1314, exemplary of the post-primitive Church accretions ceaselessly attacked by Calvinist reformers because it had no Gospel grounding, no authority preceding from Christ's record. Together with this literal defilement come the very hymns and anthems – 'but what comes out of the mouth proceeds from the heart' – Viret is careful to enumerate – sung in expiation for the desecration of the 'god of flour,' the aborted theophagy regurgitated on the monastery floor. False doctrine leading to corruption, both literal and figurative vomit, the undigested Host, the undigested, i.e., misunderstood Christ, spit up or preached by a defiled and defiling Catholic clergy to a fatally duped laity, who take in, i.e., consume, this corruption, and are taken in, that is, consumed, by the voracious defenders of a parasitical institution. Having passed out of the body, as Christ warns, through a potentially more significant orifice than the anus, it is an incarnation, so to speak, of the far worse figurative excrement of the parable: what comes out of the mouths (*de ore*) of men.

De ore, literal and Catholic, equals the doctrine of the Real Presence. Contrary to Christ's parable because contrary to Christ's intent, what goes into the mouth in this anecdote *can* defile the man, if it is theophageous rite, if it is the literal consumption of a corporeal body. The literal reading of Christ's words at the Last Supper, rather than ensure literal, corporeal salvation (i.e., health) results in vile regurgitation and implied perdition because 'what comes out of the mouth has its origins in the heart; and that is what defiles a person.' *De ore*, figurative and Calvinist, equals the rejection of literal ingestion to be replaced by the figurative dissemination and consumption of true and pure Gospel teaching. Viret and others argue for the figurative reading of the Eucharist, citing other Old and New Testament examples of the use of metaphorical speech as precedent, most frequently Christ's own standard recourse to parable. I choose an illustrative passage, almost at random, from an anonymous but suspiciously Lyonnais and 'Viretien' work (Lestringant 25), the *Sommaire recueil des signes sacrez, sacrifices, et sacremens institutez de Dieu depuis la creation du monde* (*Succinct Compendium of Sacred Signs, Sacrifices and Sacraments Instituted by God since the Creation of the World*), 1561:

> Jesus Christ said, 'This is my body'; he also says, 'The bread of God is he which cometh down from heaven.' In both these passages is not the word 'is' used? Must we necessarily infer the magic of transubstantiation rather than sanely interpret this by a use of metonymy and familiar comparison from bread to Jesus, to make us understand that by him we will be given eternal life, by him we will be given spiritual nourishment, just as by bread are we given bodily nourishment?... Why, O mass-lovers, have you undertaken to invent another interpretation than that of Jesus Christ, who witnesses that the flesh cannot profit us but that the Spirit can give life, and that his words are not carnal but spiritual, giving spirit and life by faith and assurance?.... (111–13)

> Iesus Christ dit, Hoc est corpus meum [Mark 14.22, Luke 22.19]: il dit aussi de luy, Hic est panis, qui de cælo descendit [John 6.33]. En tous les deux passages ce mot, Est, y est il pas employé? Faut il pourtant inferer une magie de transubstantiation, au lieu de sainement interpreter par une metonimie & comparaison familiere du pain à Iesuchrist, pour nous faire entendre par luy nous estre donnée la vie eternelle, & par luy nous estre donné nostre nourriture spirituelle, ainsi que par le pain la nourriture corporelle? .... Pourquoy, ô messalians, avez vous entre-

> prins d'inventer une autre intrepretation que celle de Iesuchrist, qui atteste la chair ne profiter: mais l'Esprit vivifier? & que ses paroles ne sont charnelles, mais spirituelles: donnant esprit & vie par foy & asseurance....

Thus Viret resorts to the most 'carnal' of anecdotes, uses scatology as trope, vomit as persuasive rhetorical figure, precisely because it so accurately embodies the literal, the corporeal and the terrestrial, and identified with Catholic doctrine and practice on the Eucharist, usefully labels it the categorical opposite of both Calvinist doctrine and practice, i.e., the true meaning of Christ's words at the Last Supper. And he uses scatology as trope, vomit as persuasive rhetorical figure, precisely because it so adroitly exploits the semantic playing field of the mass as food, corporeal and spiritual. 'Nourishment there is indeed in the Calvinist Communion but it's a spiritual nourishment that sustains only the soul, a celestial food that assumes on the part of he who receives it a living faith granted by grace, indispensable for achieving eternal life' (Lestringant 29).

The learned among Viret's audience could naturally have been counted upon to make these more elaborate associations. But even the 'simple folk' would have been familiar with vernacular homily of an orthodox tradition long pre-dating its co-option by Evangelicals and Reformers, and could have been expected to recognize (and perhaps be persuaded by) an argument using such basic images.[15] Images are what Viret's colorful anecdote evokes and is designed to evoke as part of a vivid discourse intended not just to debase the central Catholic rite by dragging it through the gut, but to 'explain' it in terms easily grasped. The scatological baseness of the anecdote plays out visually the material, physical farce that Viret sees in the mass: what happens to the 'objective presence of the divine' when it must past through the stomach? He expects to make his audience share his vision and react with repugnance, not to his use of such low style but to what such low style allows him to render explicit.

Viret was not alone, of course, in taking on the Roman Missal. Agostino Mainardo's notorious and notoriously abusive *Anatomia della messa* was published in 1552. Translated into French and printed in Geneva in 1555 (the year after Viret's first edition of the *Cauteles*), again in 1561 (which year also saw a Latin edition) and yet again in 1562, the *Anatomy* prefigures Viret's approach by citing and satirizing the various parts of the Roman rite and by resorting to scatology to do so. By the already familiar method of comparison with Scripture, Mainardo proposes to show up the mass for the stumbling block to salvation and 'epouvantable monstre' ('frightful monster') that he thought it to be. Chronology, geography and number of editions might encourage us to believe that Viret had knowledge of Mainardo's satire. Be that as it may, his prose text, much as the closing sonnet below, speak to the ubiquity and believed efficacy of resorting to scatological rhetoric to package reform polemic.

> Toy qui n'as point encores seu
> Qui a l'orde Messe vomie,
> Regarde ceste Anatomie:

Afin de n'estre plus deceu.
Car le monde a long temps receu
La puante, comme s'amie
Ne cognoissant son infamie,
Qu'il a maintenant apperceu.
Comme donc elle t'a seduit.
Descouvre la moy par deduit:
Afin qu'à tous soit mieux cogneue.
Mais si elle leur pue trop fort,
Renvoye-la, pour un renfort,
En enfer, dont elle est venue. (Ee^{vii})

You who have not yet seen
Who has vomited the filthy Mass,
Read this Anatomy
So as to have your eyes opened.
For the world has long accepted
The stinking thing as its lover
Not realizing her infamy
That the world has only now perceived.
Thus since she has seduced you
Uncover her infamy with delight
So that it may be seen by all.
But if she stinks too strongly for you
Send her back for help
To Hell, whence she came.

Notes

[1] English translations throughout are my own unless otherwise indicated.

[2] Pierre Viret, *Des actes des vrais successeurs de Jésus-Christ et des ses apostres et des apostats de l'Eglise papale* (1554). The full French passage below is excerpted from Chapter XXI, 'Des Dieux de la Messe, douteux & vomis' ('Concerning the uncertain and vomited gods of the Mass') of the 1563 re-issue of the missal translation and commentary, *Les cauteles et canon de la messe* 51–4: 'Je raconteray une histoire, pour l'exposition tant de ceste Cautele, que des autres suivantes, qui est advenue à Auxerre, l'an mille cinq cens vingtsix, le iour de la feste, qu'ils appellent Du corps de Dieu : laquelle il est question. Il y avoit un vieil Iocopin au Convent des Iocopins dudit lieu, qui estoit presque tout pourry de verole, en sorte qu'il tomboit presques tout par pieces. Ie me deporte de le nommer, combien que ie scache son nom & surnom. Ce bon vieil Pater, ayant esté longtemps sans dire Messe & sans faire & manger des dieux, à cause de sa vieillesse & debilité, voulant honnorer son bon dieu de farine le iour de sa feste, se mit à dire Messe. Mais il n'avoit pas estomac assés fort pour porter & digerer ce corps de son Dieu qu'il avoit mangé, ne ce sang qu'il en avoit beu. Parquoy,

comme ils s'en pensoit re-tourner en sa chambre, en passant par le cloistre du Convent, il luy eschappa de rendre sa gorge devant la porte du chapitre. Ie vous laisse à penser quelles drogues il vuida de la boutique de ce pouvre estomach tout pourry de verole, & comment ce dieu qu'il avoit mangé, fut là accoustré. Or il y eu incontinent un bruit & un trouble fort grand par tout le Convent. Et pource que le Prieur n'y estoit pas, à cause qu'il estoit allé prescher quelque part là au pres, tous les freres furent encore plus esperdus & despourveus de conseil. Toutesfois ils adviserent encores, apres qu'ils eurent long temps disputé de ceste matiere, de mettre dessus ce sainct corps, le tabernacle qu'ils avoyent accoustumé de mettre sur les sepulchres, quand ils chantent pour les trespassés. Ce qui fut fait, à fin que quelcun ne passast dessus ce vomissement sacré, ou que les chiens ne le vinsent leicher. Car c'estoit viande propre pour leur en faire banquet. Et pour honnorer d'avantage ce sainct vomissement on mit quatre cierges allumés aux quatre coings de ce tabernacle. Puis fut commandé aux novices de chanter tout le iour ce verset de l'hymne de celle feste, qui se commence, *Tantum ergo sacramentum, veneremur cernui &c.* C'est à dire, Honnorons donc estants enclinés & courbés en terre, ce sacrement si grand: c'estoit à dire pour lors, ce vomissement de monsieur le beau pere, qui estoit un sacrement fort noble & fort excellent. Ils chantoyent ausssi l'antien du mesme iour, qui se commence, *O sacrum convivium, &c.* c'est à dire, O le sacré banquet. Et non sans cause. Car il estoit bien digne de ceux qui luy faisoyent tant d'honneur. Mais cependant il ne se trouva toutesfois nul d'entr'eux, qui en vousist oncques gouster, quelque sacré qu'il fust. Car apres qu'il fut ainsi honnoré, il fut bien mis en deliberation que quelque prestre ou diacre d'entr'eux, fourrast ce sacré banquet en sa gorge, comme la bonne coustume le porte, & les Cauteles, que nous exposons à present, l'enseignent. Mais il n'y en eut pas un seul si hardy, ne si arrogant en toute la bende, qui l'osast entreprendre, & qui s'en reputast digne: tellement que iamais ils ne dirent tous de meilleur cœur, *Domine, non sum dignus ut intres sub tectum meum,* quand ils veulent manger leur dieu, qu'ils le dirent alors: c'est à dire, Seigneur, ie ne suis pas digne que tu entres sous mon toict. Brief, il n'y eut personne d'entr'eux qui le vousist loger sous son toict: c'est à dire, l'avaller & loger en son estomac. Car voila le sens, auquel ces blasphemateurs appliquent ces sainctes paroles de ce bon Centenier, qui s'adressa à Iesus Christ [Matt. 8, Luc 5.]. Puis donc que tous en faisoyent si grand refus, il fallut prendre autre conseil: c'est à dire, que ce sainct vomissement seroit levé & recueilly reveremment, avec le dieu de la boitte, & que la terre sur laquelle il avoit reposée, seroit diligemment raclee, & qu'il seroit transporté au temple. Cela conclu, le Supprieur (du nom duquel ie me deporte, combien que ie le scay) se revestit soudain des vestemens sacrés: puis tous les freres du Convent y allerent en belle procession, avec deux novices qui portoyent les chandeliers & les cierges devant ce sainct reliquaire. Les novices sont par deçà, qui en peuvent tesmoigner. Et ne faut estre esbahy s'ils ont renoncé à une telle religion, & à une telle doctrine, de laquelle ils ont veu telle practique. Car qui peut ouyr & penser ces choses, sans grand horreur? Mais qui n'est esbahy comment tout le peuple qui se dit Chrestien, n'est abysmé dix mille fois, entre lequel tels blasphemes & abominations ont duré si long temps, voire entre ceux qui ont esté estimés les colomnes de l'Eglise? Car nous voyons icy par ces Cauteles, que la practique en a este commune: voire si commune, qu'elle a esté redigee par escrit, comme nous le voyons icy. Qui eust iamais pu penser ne croire, que le peuple Chrestien fust tombé ie ne dy pas en telle brutalité, mais en telle rage & fureur? Car qu'est ce rage & fureur, si cecy ne l'est? N'y a il pas grand raison de persecuter & de brusler ceux qui ne veulent adiouster foy à ces tant enormes abominations? & qui ne veulent adorer le vomissement de ces puans poacres? O bonté de Dieu, comment as tu esté si patiente, d'endurer des blasphemes si horribles, contre la saincte maiesté de ton Fils Iesus Christ? Mais puis que ces abominables ne se contentent point seulement de ne recognoistre leurs fautes, ains qui pis est, au lieu qu'ils auroyent merité d'estre escorchés ou rostis tout vifs eux mesmes, ils perecutent & bruslement ceux qui ne veulent consentir à telles enormités: ne sont ils pas bien dignes qu'on les manifeste tels qu'ils sont, voire par leur livres & leurs faicts mesmes? Allez mainteant, messieurs les Iacopins, faire brusler ceux qui ne voudront adorer les dieux, lesquels vous & vos semblables vomissez par vostre puante gorge, & qui ne les voudront honnorer par processions avec vous. Mais achevons ceste belle histoire. Apres que monsieur le Supperieur tout revestu, & tout le Convent, fut là disposé en belle procession, la terre fut raclée diligemment, & le sacré vomissement recueilly des beaux doigts sacrés & enhuilés. Alors Iacopins se mirent derechef les uns à chanter, & les autres qui estoyent des plus bigots, à pleurer, voyant un spectacle si triste & pitoyable. Car cest inconvenient leur troubla leur feste. Apres que ce sainct vomissement fut porté dedans le temple en procession, il fallut derechef deliberer qu'on en feroit. Les anciens docteurs du Convent furent d'advis, qu'on le mist au feu, & que les cendres fussent mises & gardees au reliquaire. Car ils iugerent qu'il valloit mieux tantost ainsi faire, que le laisser pourrir & corrompre d'avantage. En quoy ils ont fort bien observé & practiqué ce qui est proposé par cette Cautele. Un chacun pourra maintenant iuger du reste, par ceste histoire, & penser avec soy mesme s'il y a bien de l'ordure, & de la puantise en ceste religion Romaine.'

[3] To list only a representative sampling of recent work in English: Carlos Eire, *War Against the Idols:The Reformation of* Worship (1986); Peter Matheson, *The Rhetoric of the Reformation* (1998); Mark Edwards, *Printing, Propaganda, and Martin Luther* (1994); and R.W. Scribner's influential *For the Sake of Simple Folk* (1981, revised and updated 1994); on the Eucharist and the Calvinist tradition in particular, see Frank Lestringant, *Une sainte horreur* (1996) and Christopher Elwood, *The Body Broken* (1999); for the most recent work on the Catholic response, see Luc Racaut, *Hatred in Print: Catholic Progaganda and Protestant Identity during the French Wars of Religion* (2002).

[4] Cf. Phil. 3.19: *quorum deus venter.*

[5] Calvin's original Latin text runs as follows: *Cur tanta saevitia et acerbitate pro missa, purgatorio, peregrinationibus, et id genus nugis belligerantur,... Cur, nisi qua illis Deus venter est, culina religio, quibus sublatis, non modo Christianos, sed ne homines quidem futuros se credunt? Tametsi enim alii splendide se ingurgitant, alii tenuibus crustulis victitant, omnes tamen ex eadem vivunt olla, quae sine illis fomentis non frigeret modo, sed penitus conglaciaraet. Ideo, ut quisque eorum pro ventre est maxime sollicitus, ita pro sua fide bellator*

acerrimus. Denique huc ad unum omnes incumbunt, vel ut regnum incolume, vel ut ventrem confertum retineant (*Opera* 1.14).

[6] In the Latin original: *Baptismata et purificationes suam illis immunditiem, spurcitiam, pollutionem exponebant, qua in sua natura contaminati erant; aliud vero lavacrum pollicebantur, quo omnes extergerentur et abluerentur eorum sordes (Hebr.9.1 Ioan.1.Apoc.1). Et hoc lavacrum Christus erat, cuius sanguine abluti, cuius livore sanati sumus (Ies.53.1 Petr.2). Sacrificia illos suae inequitatis arguebant, simulque docebant necessariam esse aliquam satisfactionem, quae iustitiae Dei solveretur* (*Opera* 1.109).

[7] Calvin's original Latin text of 1550 which he translated the same year: ... *ut de prima classe disserere exordiar, male habet fastuosos homines, et pompis deditos, quod plebeio et abiecto dicendi genere, spiritus utitur in scripturis: qui etiam ele-gantiae assueti sunt ac nitori, incultam istam et nullis coloribus vestitam oratio-nem vel respuunt, vel fastidiunt.... quin potius Ieremiae stylus oppidanum homi-nem, Amos vero pecuarium redolet: libenter fateor sacros libros, quibus tota coeli-stis philosophia continetur, non rhetorum modo picturis, sed mediocri etiam ornatu, qualem etiam literati homines desiderant, esse vacuos:...* (*Opera* 8.14).

[8] To cite but one Catholic response among many, the prolific Artus Désiré's Catholic protagonist of the 1550 *Les combatz du fidelle Papiste pelerin romain, contre l'apostat Antipapiste...* (*Battles of the Faithful Roman Catholic Pilgrim Against the Anti-Papist Apostate*), scandalized that Lutherans make Holy Writ so readily available to all and sundry, cautions: 'Dont pour demonstrer sa science / Et son esprit à mal induit, / La saincte escripture traduit / En poursuyvant sa grand' follie / Et la si farcie et remplie / De choses à la chair plaisantes, / Qu'elles sont assez suffisantes / Pour perdre les troys parts du monde...' ('And so in order to show off his knowledge / And his malignant spirit, / [Luther] translates Holy Scripture / Bent on his great folly, / And so stuffed and filled it / With things so tempting to the flesh, / That they are sufficient / To damn three-fourths of the world...') (11^v).

[9] See also G.J.C. Snoek's recent examination of the links between the cult of relics and the cult of the Eucharist in the late medieval church, *medieval Piety from Relics to the Eucharist* (1995). The standard Calvinist practice of conflating the two – the Host as relic, capable as cultic object of effecting miracles – seems to demonstrate on Viret's part an awareness of the historical process Snoek elucidates, i.e., that Eucharistic devotion was not only a recent, post-Apostolic and hence inauthentic innovation but also that it developed in imitation, conscious or otherwise, of the far older, yet easily condemned as idolatrous, practice of relic veneration.

[10] It is not my design, as I hope is clear, to argue for or against the Reformers' willful and calculated reduction of the subtleties of the doctrine of transubstantiation and thus to replay one of the major doctrinal battles of the sixteenth century, and I will not attempt to assess the distortion effected by Viret's polemic in particular. Rather am I interested in explicating his adept exploitation of the rhetorical figure or trope of scatology, whatever its merits, paying close attention to one isolated example.

[11] See Frank Lestringant's careful reading of this, what he calls the Reformers' 'recourse to grammar' (*Une sainte horreur* 15–25).

[12] On the publication history of the *Cauteles*, see Barnaud (603). Among contemporary accounts of Bèze's famous formulation expressed at the Colloque de Poissy ('Nous disons que son corps est esloigné du pain & du vin, autant que le plus hault Ciel est esloigné de la terre...,') see Pierre de La Place's 1565 account, *Commentaires de l'estat de la religion & la republique*, (254^{v}); and the rebuttal of Charles de Guise, the Cardinal de Lorraine, *L'oraison de monseigneur le illustrissime & reverendissime Cardinal de Lorraine*, as well as the reaffirmation of the thirteenth session of the Council of Trent, (11 October, 1551).

[13] The Latin text reads: *Item si hostia, vel pars hostiae inventa fuerit sub palla vel corporali, & dubitatur si est consecrata vel non, debet eam post sumptionem sanguinis reverenter sumere: ut in titulo, De celebratione Missæ plenius invenies. Item, si corpus Domini infirmo datum, propter infirmitatem, sive per aliam qualemcunque causam reiicitur: prout deligentius fieri poeterit, recipiatur: & si propter humanam fragilitatem invenitur reiectionem illam velit sumere, aut non possit, coburateur: & combustio in sacrario reponatur.* (645–47)

[14] From the *Decree on the Holy Eucharist*, ch. 5: 'The holy council furthermore declares that the custom which has grown up in the Church of God of honoring this great and adorable Sacrament with special worship and solemnity on a particular feast day each year, and the custom of carrying the Sacrament in procession through the streets and public places with reverence and respect, are good and very religious customs.'

[15] On the importance of signs, portents and images, figured in both natural and supernatural events, to the popular imagination in sixteenth-century French religious and civil controversies, see both Denis Crouzet's magisterial two-volume study, *Les Guerriers de Dieu* (1990) and Larissa Taylor's *Soldiers of Christ* (1992).

Chapter 5

Community, Commodities and Commodes in the French *Nouvelle*

Emily E. Thompson

As the expression, 'être pressé de son honneur' ('to have one's honor put under pressure, to feel the need to urinate or defecate') (*Dictionnaire du moyen français* 350) attests, sixteenth-century men and women in France associated defecation and (dis)honor. Honor could be sullied ('honnir') and a single word referred to those without honor and without cleanliness – 'vilain.' In the *Heptaméron*, Marguerite de Navarre regularly characterizes her morally depraved characters as 'ord et salle' ('vile and dirty'). Since Renaissance honor was inextricably linked to class, the notion of excrement evoked not only the dishonorable, but also the lower classes as Mikhail Bakhtin has so famously demonstrated. In reality as well as in discourse, the social elite had a different relation to scatology (the likelihood of having a latrine installed in one's home, for example). It is important, however, not to overstate the *pudeur* of the upper classes. As Montaigne reminds us at the end of the century, 'Et les Roys et les philosophes fientent, et les dames aussi' ('Both kings and philosophers defecate, and ladies too') (1063).[1] As notions of intimacy and privacy were beginning to take form in the sixteenth century, how were scatology and scatological discourse perceived by different social classes? In order to address one small part of that question, I turn to the *nouvelle,* a particularly popular French Renaissance genre that uses the mundane as a setting. I explore the limits of social privilege in four related scatological tales from the collection of King Francis I's sister, Marguerite de Navarre, first published in 1558, and from the *Cent Nouvelles Nouvelles* (ca. 1515), written by Philippe de Vigneulles, a merchant from Metz.[2]

Before turning to the four specific stories in question, I would like briefly to compare the scatological nature of the two collections. The *CN*, reminiscent in tone of the medieval *fabliaux,* are regularly punctuated by tales that the editor Charles H. Livingston characterizes as 'écœurante[s]' ('disgusting') (308). These tales treat traditional themes where sex and defecation/urination are confused, where scatology serves to humiliate an adversary, where bodily functions provide an outlet for repressed violence and where shit retains an inherently comic value. The image of the sickly narrator in the prologue of the *CN* reinforces the disordered, unhealthy social body illustrated in the individual scatological tales. Like the narrator, confined to his bed with a diseased body, the characters, too, are restricted to intimate interior spaces where a conflation of multiple bodily functions occurs. The recurring theme of ingesting fecal matter, for example, provides a temporary relief of social ills through the inversion of top and bottom, but promises no long-term cures. What has been ingested will once again be transformed into excrement suggesting the impossibility of ever escaping from the stink of the human body or of human society.

Scatology does not figure as prominently in the *nouvelles* of the *Heptaméron*, but it is already introduced in the opening sentence through the use of the term 'fange':

> Le premier jour de septembre, que les baings des montz Pyrenées commancent d'entrer en leur vertu, se trouverent à ceulx de Cauderetz plusieurs personnes tant de France que d'Espaigne; les ungs pour boire de l'eau, les autres pour se baigner, et les autres pour prandre de la fange; qui sont choses si myraculeuses que les malades habandonnez des medecins s'en retournent tous guariz. (1)

> On the first day of September when the baths of the Pyrénées begin to assume their potency, several people, as many from France as from Spain, found themselves at those of Cauterets, some to drink the waters, others to bathe in them, the rest to have mud treatments which are remedies so miraculous that the sick abandoned by doctors return from these baths completely cured.

Although Marguerite's narrators are also telling stories to fight off threats to their health, their noble diseases and cures evoke an entirely different setting. These patients have not only left their beds and their rooms, they have left Paris to take a perilous and expensive journey to the curative waters of the Pyrénées. Here they have bathed and drunk the waters and taken the 'fange,' a term that implies moral as well as physical soiling. In Marguerite's lexicon, and specifically in the tales examined in this article, mud, excrement, and immorality are frequently brought together by another term: 'ordure.' Her collection thus begins with a curative form of excrement, one removed from the bedroom and the privy and taken into the macrocosm of France. Although the internal *fange/ordure* remains a symbol for corruption, bringing *fange/ordure* out into the open, even applying it externally, appears to have a potentially salutary effect. Because they can cure the social body, Marguerite's scatological words 'ne sont jamais puantes' ('are never smelly') (402), but instead assume the perfume of salvation. As in Vigneulles's *Cent Nouvelles Nouvelles*, the scatological metaphor is closely related to the narrative project. For Marguerite, however, verbal and physical purging remain acts of hope.

Unlike most of the scatological stories in sixteenth-century *nouvelle* collections, the first two tales I will analyze (Marguerite's tale 11 and Vigneulles's tale 80) describe excremental accidents, not pranks. Scatology functions as a means of humiliation, but it is the protagonists' own shortcomings and not a personal enemy that are responsible for this degradation. The main characters are unavoidably associated with their natural by-products through the narration of the act of defecation. The fecal matter in question thus retains its literal, biological dimension as well as performing as a symbol. The victims in both tales are of noble blood and share several weaknesses, being excessively proud, solitary and rash as well as prone to verbal exaggeration. The woman in Marguerite's tale is also prudish while the male character in Vigneulles's story has an obsession with cleanliness and a dictatorial manner with his family and servants, 'tant dangereux qu'il n'en y avoit point au pays du pareil' ('so finicky that no one else in the country could compare') (308). Their humbling occurs through their bodily functions, their exaggerated reactions to what is natural, and a humiliating exposure to their perfect Other.

Marguerite's tale places a noblewoman, Madame de Roncex, in a Franciscan monastery where she is suddenly overcome by a natural urge. She takes another woman along for company, but then makes her stay outside the privy, a decision indicative of her 'honteuse et secrette' ('modest and secretive') nature according to the narrator (108).[3] Because the privy is too dark for her to notice the disgusting by-products of the monks' latest drunken revelry, she sits in the filthiest spot and becomes literally stuck to the seat. Crying for help, she uses characteristic hyperbole and declares herself 'perdue et deshonnorée' ('fallen and dishonored') (108). The woman waiting outside jumps to the conclusion that Madame de Roncex is being raped by monks (an assumption in line with all that she has heard about the Franciscans) and seeks help among the noblemen who have accompanied them to the monastery. When they arrive on the scene, they are treated to the 'beau spectacle' ('fine show') of Madame de Roncex, bare-bottomed and wallowing in shit (109). Shamed by their presence and their laughter, she drops her skirts, thus completing the contamination of her clothes. She must eventually be completely stripped and cleaned, 'despoiller toute nue,' before leaving the monastery (109). Madame de Roncex first feels anger towards her well-meaning female companion, but is soon overcome by the crowd's merriment. When she begins to laugh with the others, she transforms their mocking laughter into communal mirth that, in turn, is echoed by the frame-narrative's group of tale-tellers or *devisants* who also 'se prin[dren]t bien fort à rire' ('started to laugh hard') (109).

The story contains two elements that strike the modern reader as illogical and that suggest the symbolic significance of this tale. First the monks' vinous excesses are transformed not into urine, but feces. Renja Salminen in her 1999 edition of the *Heptaméron* suggests that Marguerite chose to term the monks' intake as 'moust de Bacchus' ('sweet wine of Bacchus') for the play on words with 'bas cul' ('ass'), thereby reinforcing the carnivalesque nature of this story while reminding the readers of the monks' reputation for drunkenness (697). Like her contemporaries, Marguerite associated indulgence in wine with defecation.[4] Besides suggesting certain limitations in sixteenth-century medical knowledge, this association also permitted Marguerite to substitute the strongly negatively coded lexicon of 'fiens' ('filth,' 'scat') for a much less evocative description of a urine-soaked privy.[5]

The *fiens* terminology appears in her spiritual poetry in self-deprecating passages like the following from *Le Miroir de l'âme pécheresse*: '...quant a moy, je suis trop moins que riens: / Avant la vie boue, et après fyens' ('...as for me, I am too much less than nothing, / In life mud and afterwards scat') (45–6). Marguerite was certainly not the only religious writer to use the scatological image. She restricts its meaning, however, to the theme of *mortificatio* – an initial fall into sin, followed by a shame-induced purification – and finally complete union with God. The tale illustrates the *mortificatio* in a very simple allegory. The fall into sin is here quite literally a lowering into the abasing fecal matter. The ensuing *purgatio* strips the subject of the clothes that she uses to hide her 'honteuse et secrette' body, leaving only her vulnerable and cleansed nudity. Finally, her ability to see her predicament from the point of view of the others and to adopt their laughter suggests the abandonment of her body and her individuality and a perfect union with what was formerly exterior to her. Here the juxtaposition internal/external attributes a negative value to the inside of the latrine recalling Marguerite's use of the prison image in her poetry.[6]

That Madame de Roncex's religiously connoted shame is purged of the foul odors of scatological comedy is evident from the absence of any reference to senses other than that of sight. This brings me to the second, illogical moment of the story. The men who rush into the dark privy to save Madame de Roncex witness the 'beau spectacle' of her exposed and soiled rear end even though she herself was initially unable visually to assess the state of the latrine. The narrator who announces the following tale promises to treat the topic of 'amour aveugle' ('blind love') (110). Madame de Roncex, too, apparently suffers from a metaphorical blindness. This, then, helps to explain pious Oisille's rather startling comment during the discussion that follows Nomerfide's dirty story: 'Combien que le compte soyt ord et salle, congnoissant les personnes à qui il est advenu, on ne le sçauroict trouver fascheulx. Mais j'eusse voulu veoir la myne de la Mothe et de celle à qui elle avoit amené si bon secours! ('Although the tale is filthy and dirty, knowing the people to whom these things happened as we do, we could not find it upsetting. But I would have liked to see la Mothe's expression and that of the woman to whom she brought such good succor') (109). The fact that Oisille would have wished to view the humiliated Madame de Roncex herself might at first seem odd. But it is not Madame de Roncex's posterior that Oisille would have liked to see, but her 'myne' ('expression') as she received succor. It is not so much sight, but spiritual *in*sight that Oisille would have liked to share with the joyous crowd. Even the ill-reputed monks seem to be exonerated and allowed into this larger community. Excrement in tale 11 thus reduces the female character to her besmirched body in order to teach her humility and moderation. Once the dirty is cleansed and anger and humiliation are transformed into laughter and forgiveness, communal joy can replace the solitary retreat.

At the beginning of Vigneulles's story 80, the narrator sets a tone that resembles the project we have just seen in the *Heptaméron*. He promises a merry tale that will amuse the entire company, each listener will have 'sa livrée' ('his/her regalement') (308). The tale ends, however, without any sign of mirth on the part of the narrator or the characters and without any examples of equity or community. The story contains elements that might also imply spiritual impurity. For example, the principal character demonstrates asocial behavior, attributed to 'ygnorance ou follie' ('ignorance or folly'), by refusing to share a communal cup of wine (308). His symbolic rejection of the Catholic community is a function of his pride and obsessive cleanliness that strike the narrator as 'chose de l'autre monde' ('otherworldliness') (309). Like Madame de Roncex before she enters the privy, this character also fails to recognize that he is very much part of this imperfect physical world. Unlike Madame de Roncex, however, the noble protagonist of this tale is never named and thus never nominally or literally exposed. The reader might judge this tale 'fascheux' ('upsetting'), then, referring once again to Oisille's standards: 'Combien que le compte soyt ord et salle, congnoissant les personnes à qui il est advenu, on ne le sçauroit trouver fascheux' ('Although the tale is filthy and dirty, knowing the people to whom these things happened as we do, we could not find it upsetting') (*Heptaméron* 109). And, indeed, just as Vigneulles's character never achieves true mortification, the lesson in this *nouvelle* likewise falls short of Marguerite's Evangelical message.

The narrator instead situates the protagonist in a distinctly social context. He is 'bien fondez et rantez' ('well established and landed'), and 'miste et honneste en abit et au boire et manger' ('elegant and refined in apparel and in food and drink'); he enjoys the hunt (308). In fact his fault is stated thus: 'plus honneste estoit qu'il ne luy fut besoing' ('he was more respectable than he needed to be') (308). 'Honneste' ('respectable') refers first to his standards of cleanliness and propriety, but also, unavoidably, to class. As Elias, Muchembled, Berrong, and others have noted, the changing levels of tolerance towards the lower part of the body and all its functions are closely linked to conceptions of class difference in sixteenth-century Europe. Is Vigneulles's character, then, also too much of a noble? His fastidiousness certainly makes trouble for another class, his servants, who are held to the same impossible hygienic standards, increasing their labor without commensurate gratitude or compensation.

The details of the plot, like tale 11 of the *Heptaméron*, focus on the relationship between the noble character and a single servant. The unnamed, maniacally clean nobleman has just overindulged in drink (like the monks in Marguerite's tale) and feels that equalizing of all urges, the 'besoing a chascune creature de fere sa necessitté' ('the call of nature') (309). After relieving himself, he reaches for some straw or a rag. Less learned in *torcheculs* than Gargantua, he accidentally sullies one of his fingers. Again, disgust translates into cries of distress that lead a servant to suspect a far graver emergency.

In this case, however, the misunderstanding is immediately clarified. Before any others arrive on the scene to witness the protagonist's humiliation, he leaves the confining space of the *cabinet*, but experiences no epiphany in crossing this threshold. He refuses to let his overall authority suffer an indignation he confines to the contaminated digit, so he demands that his servant cut it off. The servant, a man of good sense, tries unsuccessfully to persuade his master that the stigma is but superficial: 'esse tout? … ne le vous fault rien que laver, et il s'en yra' ('is that all? …for that, all you need is to wash and it will go away') (309).[7] Faced with his master's intransigence, he fetches an axe, but decides on a more humane compromise, bringing it down blunt end first on the master's finger. The latter, once again betrayed by his own body, responds instinctively and trying to calm the pain, puts his finger into his mouth. This added humiliation provokes a murderous rage in the master who only returns to reason once the effects of the wine finally wear off. In the end, however, he is said to appreciate the servant more for having intervened so wisely.

The nobleman's humbling has a single witness, and the rest of the estranged community (wife and other servants) do not join in merriment. Although his anger dissipates, the nobleman emits no selfless laughter at the end. The servant may have derived a private chuckle recounting the tale to the narrator, but is never allowed an empowering, public laugh. Not even the narrator seems amused; instead he transforms the conclusion into a warning against other proud, unnatural men:

> Toutefois, comment qu'il en fut, luy qui estoit le plus dangereux du monde bouta son doigt breneu en sa bouche de haiste qu'il eust pour la doulleur qu'il sentist, de quoy il en eust les dentz deschiez. Et ainsi en preigne a tous ceulx qui le ressemble. (310)

> All the same, no matter how things ended, he who was the most finicky man in the world, because of the pain he felt, stuck his crap-covered finger into his mouth and as a result got shit on his teeth. And may the same befall all those who resemble him.

The use of 'toutefois' ('all the same') reinforces the absence of a satisfying conclusion to this tale of folly and injustice. The master, inferior as he is in common sense and humanity, retains his authority. No symbolic emasculation occurs even though the servant is in a position to assert himself. And although this threat of violence is averted, class tensions are maintained. The narrator never suggests an evolution of the master's behavior or self-consciousness. Although the reader can assume that the nobleman has cleaned up in a literal sense at least (the finger supposedly 'retourna à garison' ['returned to health']), the last image of the *nouvelle* is that of the nobleman's shit-covered teeth (310). In this story, Vigneulles could be evoking the expression, 'S'en laver les dents' ('to wash one's teeth of something') (*Dictionnaire du moyen français* 185), implying a disinterested attitude like the one Madame de Roncex manages to assume at the end of her tale of mortification. In contrast, the unnamed nobleman is left with his teeth stained, still stuck inside the mentality of the introspective 'retraict' ('privy').

This general image recurs frequently in the *CN*. The theme of excrement that finds its way onto characters' heads and/or into their mouths links 6 out of the 7 scatological tales in the collection. Besides an obvious allusion to a world turned topsy-turvy, the image of swallowing shit is the ultimate representation of abasement. 'Avaler' refers to a falling, or lowering, *aval* ('downstream') versus *amont* ('upstream'). Lutheran propaganda, for example, shows Luther's enemy Johann Cochleus swallowing the Devil's shit and then defecating his insults against Luther (Scribner 85, figure 62). In a secular context, too, excrement can evoke a violent contempt and a threat of disorder. For Vigneulles's nobleman, human waste does not simply remind him of his innate worthlessness before God, but suggests a contamination that could threaten his privileged status here on earth: 'helas, je suis homme perdu,' ('Alas, I am fallen, ruined') he shouts (309), using the same polyvalent adjective as did Madame de Roncex in her cry for help.

Both the *Heptaméron* and the *CN* also treat the familiar story of the frozen turd (*Hept.* 52, *CN* 24). In these related tales, fecal matter is intentionally used to humiliate an adversary. Again, Vigneulles's story contains threats of violence and a semi-private humiliation, this time in the context of unresolved gender and age tensions. No laughter occurs on the part of the characters or the narrator despite his own role in the hostile prank. Marguerite's tale 52 repeats elements from tale 11, once again associating excrement with laughter and self-awareness.

Vigneulles's *nouvelle* 24 pits an elderly woman (who may or may not be a fishmonger) against a group of young hooligans. Trying to get rid of her nagging presence, they first expose themselves to her, then throw a frozen turd, 'un groz estron qui estoit tout engelez' (127), at her head. Mistaking it for a rock, she collects it as evidence of their assault. But, since it is late, she decides to go home and to report them to the authorities the following morning. When she awakes and goes to show the 'rock' to her husband, she finds herself with a handful of shit, 'elle se deschia les doigts' (127). Ashamed and silenced (as the young men intended her to be), she abandons her plans for revenge.

Once again a wronged character is in a position to pursue a sort of justice. However, this time it is the victim who is contaminated by the excrement. Rather than inspiring rage, her contact with the melted turd instantly dissolves her anger into shame. Whereas Madame de Roncex's anger gives way to the redemptive laughter of a community and even Vigneulles's nobleman exchanges his anger for a loving forgiveness (devoid though it may be of repentance), the elderly woman of tale 24 gains nothing in return for her rage. Instead, she resigns herself to what she already has: 'ains se tint avec ce qu'elle avoit …' (127). Reading the threat implicit in the thawing turd, she realizes that even she has property and a status to lose. This resignation has no parallels with Christian humility, for the lesson is inflicted by a self-serving narrator whose sole concern is maintaining social barriers, not surmounting them.

Marguerite's tale 52 is far longer and exists in two quite different versions.[8] In both stories a lawyer and a nobleman are outwitted by an apothecary's valet, who leads them to believe that they have stolen a loaf of sugar from him rather than a turd. Gloating in front of a tavern fire with their stolen goods, they begin to notice a stench. Confident that it must come from the tavern itself, they turn to insult the serving maid. When they realize that their sugar loaf has melted into an accusatory slime, they are finally forced to realize the valet's triumph over their greed.

The differences between the two versions of this *Heptaméron* tale focus on the lawyer, the maid, and the valet: all socially inferior to the country nobleman. In one version, the lawyer is responsible for the initial theft, motivated by his ongoing rivalry with a clever valet. In this account, he also joins the gentleman in belittling the tavern maid and later finds himself publicly ridiculed along with his noble companion. This ridicule is administered first by the tavern maid, 'courroucée dont ilz l'appelloient ainsi villaine' ('angered that they had thus called her vile') (401), then by bystanders who learn of their dishonesty. The lawyer and the nobleman do not ever laugh at themselves, but accept the mockery as a temporary defeat in an ongoing rivalry. This communal laughter has a decidedly social timbre to it, and the nobleman is as implicated as his lower class companions in an economy of exchange: of tricks, laughter, money and fecal matter.

In the other version, the lawyer becomes a passive follower, the valet is deprived of his ingenuity, and the tavern maid is never given an opportunity to avenge herself of the nobleman's unjust accusations of uncleanliness. No crowd witnesses the nobleman's humiliation, and he alone draws the lesson from it. In this version it is the nobleman's fox fur coat that is soiled, recalling Madame de Roncex's garments in tale 11. It is clear that this version privileges the nobleman's own realization of his shortcomings – a demeaning greed and a petty dishonesty. No one laughs at him, because it is not the victory of a clever valet or a sharp-tongued tavern maid, but a personal victory of self-understanding. Even the lawyer is uncharacteristically left speechless.

The discussion following both versions remains the same, once again evoking the theme of laughter. The frame-narrative *devisants* distinguish between, on the one hand, 'parolles [qui] ne puent point' ('words that don't stink') like the turd story that demonstrates a certain truth and, on the other, morally offensive 'villaines [parolles]' ('vile, base words') (402). The controversy revolves around the women's reaction to the 'villaines [parolles]'. The men accuse them of hypocrisy for laughing at words they would never stoop to utter. Parlamente claims that women laugh, as they would at any other harmless blunder, when these words are used unintentionally. But she also accepts hypocrisy from any woman who would find amusement in the deliberate use of these words, urging her to hide her laughter behind a 'touretz de nez,' a type of mask (403). Thus hiding one's nose after the tale becomes the equivalent in hypocrisy to holding one's nose within the story. The *devisants* are split as to whether or not minor sins should be hidden by 'touretz de nez' or 'manteau de vertu' ('a cloak of virtue') (404). Even though they recognize the mortal risks involved in disguising shortcomings, they nonetheless agree to protect the transgressor in the following tale by suppressing her name.

As in the case of tale 11, it is Oisille who legitimizes the dirty story. She allows Simontault to tell tale 52, no matter how 'ord et salle' it may be as long as he uses it to convey a truth. Although excrement has been brought out from the secrecy of the 'retraict' or privy, another discourse has been thrust back into a smelly, dark interior. The only true 'villaines [parolles]' in tale 52 are the words uttered by the unique 'villaine', the much maligned tavern maid, and by the valet. The second version eliminates her nasty repartees and the valet's irreverent final denunciation as well as repressing the tavern maid's mocking female laughter. In both versions the 'manteau de vertu' that is the nobleman's fox fur coat covers a stench. The difference lies in the uncovering of the impure gentleman. In one case, it is his social inferiors who publicly reveal his putrefaction, 'il n'y a merde, si vous ne la y avez apportée' ('There is no shit unless you brought it here') (401), whereas the realization is his to make and to utter in the second version. Like her *devisants*, Marguerite hesitates between recognizing a need to protect the nobleman with some degree of social hypocrisy and the desire to turn the soiled coat inside out.

Both Marguerite and Vigneulles depict human fecal matter as a humbling equalizer, yet the humiliation of Vigneulles's characters does not push them to accept the other as an equal. These characters recognize their vulnerability to the implicit threat of *ordure* – violence and chaos – but this only reinforces the social, gender and age categories that structure their secular world and to which they cling. Marguerite's tale 11 implies a more radical transcendence leading to a classless, genderless community of the merry faithful. But the two versions of her tale 52 suggest the difficulty of producing a laughter devoid of gender and class significance. As long as Marguerite imagines the transcendence of human society, excrement remains a sign of worthlessness. Within a secular context though, she, too, recognizes its corruptive potential: excrement not just as a symbol of worthlessness but as an agent that diminishes social worth in relation to others. Like Oisille, Marguerite seems to believe that scatological discourse and a higher truth can find a common ground, but in everyday life neither 'villaines [parolles]' nor 'parolles des villaines' should be listened to or spoken.

Even in fecal matters, then, Marguerite and Vigneulles ultimately concur that ladies and chambermaids are not interchangeable. Not only can the lady not send the chambermaid to take her place in the latrine ('au lieu où l'on ne peult envoyer sa chambriere'), as Nomerfide sarcastically notes in tale 11 (108), the separation of the privy proves all-important in shielding the mortal vulnerability of the lady from the gaze of her social inferior. The lady must indeed go to the latrine by herself. And the question remains as to who will eventually emerge from this intimate space: a lady or a chambermaid?

Notes

[1] All translations in the essay are my own unless otherwise noted.

[2] This text will subsequently be referred to as *CN*, using Armine Kotin's abbreviation to avoid confusion with the anonymous fifteenth-century *Cent Nouvelles Nouvelles* (*CNN*).

[3] One possible translation for 'secret' is 'privy,' thereby reinforcing the link between the character's psychological and physical retreat into solitude.

[4] I thank Alison Stewart for having brought my attention to Erhard Schön's *Four Effects of Wine* (1529). The character in Vigneulles's tale 80 also defecates in a state of inebriation. See Stewart, 'Paper Festivals and Popular Entertainment.'

[5] Far from featuring in tales of debasement, urine is most commonly associated with the medical profession (uroscopy) in French Renaissance literature (see also Joseph Tate's essay in this volume). Vigneulles does tell of a cuckolded husband who makes his rival drink his wife's urine in revenge (*CN* 15). But, in this

case, urine takes precedence over excrement as a means of humiliation because of the narrative necessity for a substance that could be confused with white wine.

[6] The juxtaposition internal/external is a common one in Marguerite's writing. The values are frequently reversed, however. In other tales and social contexts, the surface or appearance becomes the negative terms while the soul and the individual conscience assume a positive significance. See tale 21, for example.

[7] Although he attempts to downplay the implications of his master's slip, the suspicious spelling of the respectful formula the servant uses to address him, 'mon tres chier seigneur' (my dearest lord) suggests otherwise: *chier*, an alternate spelling of *cher* (dear), could be confused with the verb *chier* (to shit).

[8] The Salminen edition includes both versions of the story as well as an explanation of the manuscripts in which each can be found (767).

Chapter 6

Pissing Glass and the Body Crass: Adaptations of the Scatological in Théophile

Russell Ganim

Introduction

Examination of scatological motifs in Théophile de Viau's (1590–1626) libertine, or 'cabaret' poetry is important in terms of how the scatological contributes to the depiction of the Early Modern body in the French lyric.[1] This essay does not examine Théophile's portrait of the body strictly in terms of the 'Baroque' or the 'neo-Classical.' Rather, it argues that the scatological context in which he situates the body (either his, or those of others), reflects a keen sensibility of the body representative of the transition between these two eras. Théophile reinforces what Bernard Beugnot terms the body's inherent 'eloquence' (17), or what Patrick Dandrey describes as an innate 'textuality' in what the body 'writes' (31), and how it discloses meaning. The poet's scatological lyric, much of which was published in the *Parnasse Satyrique* of 1622, projects a different view of the body's 'eloquence' by depicting a certain realism and honesty about the body as well as the pleasure and suffering it experiences. This Baroque realism, which derives from a sense of the grotesque and the salacious, finds itself in conflict with the Classical body which is frequently characterized as elegant, adorned, and 'domesticated' (Beugnot 25). Théophile's private body is completely exposed, and, unlike the public body of the court, does not rely on masking and pretension to define itself. Mitchell Greenberg contends that the body in late sixteenth-century and early seventeenth-century French literature is often depicted in a chaotic manner because, 'the French body politic was rent by tumultuous religious and social upheavals' (62).[2] While one could argue that Théophile's portraits of a syphilis-ridden narrators are more a reflection of his personal agony rather than that of France as a whole, what emerges in Théophile is an emphasis on the movement, if not decomposition of the body.[3] Given Théophile's public persona and the satirical dimension of his work, it is difficult to imagine that the degeneration he portrays is limited only to his individual experience. On a collective level, Théophile reflects what Greenberg calls 'a continued, if skewed apprehension of the world in both its physical and metaphysical dimensions'

(62–3) typical of the era. To a large extent, the body Théophile depicts is a scatological body, one whose deterioration takes the form of waste, disease, and evacuation as represented in both the private and public domain.

Of course, one could cast aside any serious reading of Théophile's libertine verse, and virtually all of scatological literature for that matter, as an immature indulgence in the prurient. Nonetheless, it was for his dissolute behavior and his scatological poetry that Théophile was imprisoned and condemned to death. Consequently, this part of his work merits serious consideration in terms of the personal and poetic (if not occasionally political) statement it represents. With the exception of Claire Gaudiani's outstanding critical edition of Théophile's cabaret lyric, there exist no extensive studies of the poet's libertine œuvre.[4] Clearly however, these poems should be taken seriously with respect to their philosophical and aesthetic import. As a consequence, the objective becomes that of enhancing the reader's understanding of the lyric contexts in which Théophile's scatological offerings situate themselves. Structurally, the reader sees how the poet's libertine œuvre is just that – an integrated work in which the various components correspond to one another to set forth a number of approaches from which the texts are to be read. These points of view are not always consistent, and Théophile cannot be thought of as writing in a sequential manner along the lines of devotional Baroque poets such as Jean de La Ceppède and Jean de Sponde. However, there is a tendency not to read these poems in their vulgar totality, and to overlook the formal and substantive unity in this category of Théophile's work. The poet's resistance to poetic and cultural standards takes a profane, if not pornographic form because it seeks to disgust and arouse while denigrating the self, the lyric other, and the reader. Théophile's pornography makes no distinction between the erotic and scatological. The poet conflates sex and shit because they present a double form of protest to artistic and social decency while titillating and attacking the reader's sensibilities. Examination of the repugnant gives way to a cathartic experience which yields an understanding of, if not ironic delight in, one's own filthy nature.

The Lighter Side of Libertinage: 'A un marquis'

Initial analysis of Théophile's cabaret lyric requires a more precise definition of what is meant by the 'scatological' and the 'libertine.' In the strict physical sense of the term, the scatological refers to excrement, particularly urine and feces. However, human waste also includes various bodily emanations such as mucous, genital discharge, and sweat. For Théophile, the focus is on urine, as well as these non-fecal emissions, all of which are by-products of his ongoing battle with syphilis. Théophile's fixation on the anus also reinforces the scatological nature of certain poems. Such a dimension is further developed by the idea that from an aesthetic standpoint, scatology denotes what the *Oxford English Dictionary* terms as 'filthy literature.' Through this description, readers establish the link between the scatological and the libertine. Certainly, many definitions of the word *libertin* existed in the early seventeenth century, but most dealt with sexual freedom and/or atheism (Perceau 11). The raw sexuality and physicality of Théophile's verse, which led to his persecution by the Jesuits, clearly engendered the view that this part of the poet's œuvre corrupted both social and moral order.

Despite the agony Théophile endured by his own hand through venereal disease, and by the hand of the Church in the form of condemnation, the inherently hedonistic aspects of Théophile's life and letters play a major role in setting the tone of his cabaret poetry. One work which stands out as an unabashed celebration of decadence is the poem 'A un marquis,' where the first quatrain in particular highlights the spirit of libertinage:

> Marquis, comment te portes-tu?
> Comme quoy passes-tu la vie?
> Si tu n'as d'aujourd'hui foutu
> Ces vers t'en donneront envie. (1–4)
>
> Marquis, how do you carry yourself?
> As what do you spend your life?
> If you haven't fucked today
> These verses will make you want to.

Théophile's intense affinity for the Marquis suggests a homosexual undercurrent which is reinforced in verse two, where the poet asks how, or more precisely as what ('comme quoy') the Marquis spends his life. This aspect of Théophile's affection remains largely unformed throughout the poem, and in the cabaret œuvre in general. Yet, given the poet's admiration of the Marquis, such a question implies that the young nobleman, and quite likely Théophile himself, are disposed toward sex with either men or women. Implicit homosexuality becomes part of Théophile's libertine construct. Théophile defines the Marquis exclusively as a sexual being, and in verse four intimates that the poem is conceived in part to elicit the latter's desire. Unlike some of his darker lyric offerings, the poet deals only with the pleasurable aspects of sexuality in this work. There is no lamenting of a fallen state, nor is there mourning of past escapades. If indeed the Marquis experiences unfortunate consequences because of his behavior, there is every likelihood he will recover. As the poet remarks in the next strophe:

> Es-tu gaillard? Es-tu dispos?
> T'apperçois-tu que tu guerisses?
> Ce couillon est-il plus si gros?
> Sens-tu du mal lorsque tu pisses? (5–8)
>
> Are you randy? Are you in good form?
> Do you think you're cured?
> Is your nut still as big?
> Does it hurt when you piss?

The painful urination associated with syphilis becomes only a minor setback, as the scatological reference carries only humorous, rather than lethal, connotations. Most assuredly, the Marquis will heal, the swelling around his testicles will go down, and the burn will dissipate so that the young nobleman 'si amoureux de la desbauche' ('so in love with debauchery') can continue his exploits. Scatological references gain importance in this poem because they inform the reader about the state of the Marquis's penis, which in turn serves as a gauge of the young nobleman's sexual, and thereby physical, health. Searing evacuation clearly hinders the ability to be as 'gaillard' ('randy') and 'dispos' ('in good form') as one would like. But the movement from the temporarily sickened penis and testicles in the second stanza to what the poet describes as 'Les nerfs souples et le vit roide' ('Supple nerves and a stiff cock') in the final verse underscores the re-establishment of medical and sexual order. Figuratively speaking, what occurs over the course of the poem is that the erotic symbol of healthy masculinity, i.e., the erect phallus, triumphs over the scatological symbol of the diseased genitalia. As in other poems, the scatological in this work signals illness, but unlike other texts where the poet or his addressee is overcome by sickness, here the disease is overcome by the Marquis's youth and general vigor.

Somewhat ironically, the understanding of Théophile's use of sexual and scatological contexts is deepened by the way in which the poet combines these motifs with devotional themes. As Martine Debaisieux has shown in her work on Jean Auvray (c. 1580–1630), satirical and religious traditions of the late sixteenth and early seventeenth centuries exercised a reciprocal effect on one another.[5] Of note is how Théophile incorporates God into a poem about urine, sexual debauchery, and venereal disease. As if to substantiate, in a literary manner, his concluding image of the erect member, Théophile mentions that tales of the Marquis's adventures will resound long after the nobleman's death: 'Selon le cours de la nature, / Ton esprit au lieu du bordeau / Discourra de la sépulture' ('As is nature of things, / Your vigor in the brothel / Will tell its tale beyond the grave') (13–15). On one level, the idea that an erotic narrative will emanate from the grave itself can be seen merely as a satirical device meant to introduce blasphemous discourse into an already obscene text. Yet, given the lack of bitterness and invective seen in other of Théophile's poems, the reader has license to interpret these references to the supernatural as a divine sanction of human sexuality for its own sake. Visits to the whorehouse become the stuff of epitaphs because God himself wants humanity to view sex not as diseased, i.e., as scatological, but as healthy.

In his final stanza, Théophile states:

Mais que sert-il tant de resver
En méditation si froide,
Tant que *Dieu* nous veut conserver
Les nerfs souples et le vit roide? (16–20)

What good is it to dream
In such cold meditation
If *God* wants us to keep

Supple nerves and a hard cock?

Contrary to Théophile's depiction of God in his more lugubrious poems, Théophile's view of the divine seems favorable here. In verses 19 and 20, the poet suggests that God's will is to keep humanity erotically fit and active. Théophile establishes a contrast between divine nature – whose design is that humankind enjoy sex – and human nature, which passes its time uselessly 'dreaming' and 'coldly meditating' on sexual matters rather than engaging in them. Without question, Théophile's 'theology' comes across as intentionally mordant and sacrilegious. But at the same time, this representation of God suggests a fun-loving, libertine deity created in Théophile's own image. The poet sees God as an entity who allows humanity to benefit from Creation's endowment. Conversely, humankind establishes barriers to this natural merriment through the Church, and its endless efforts to impose restrictions on human sexuality. A more sympathetic vision of God changes the role of the scatological and the venereal.

Certainly, the illness signified by burning urine reminds humankind of its fallibility. However, at least in this particular text, the malady is not a malediction that represents God's cruelty or indifference. God and nature may conjoin sexuality and disease, but the combination is not necessarily destructive. In an effort to synthesize elements that seem grossly incompatible, Théophile depicts a world in which open sexuality, venereal disease, and divine will co-exist – if not peacefully, then at least without threatening one another. The scatological enhances the libertine component of the text by evoking surprise (if not shock), sensuality, and a considerable measure of the grotesque. Within the corpus of Théophile's cabaret poetry, this text outlines the major tonal and thematic constituents that the poet will try to reconcile, with varying degrees of success, as he sets forth his view of the body and libertinage..

Misogyny and Masochism: 'Stances contre une vieille'

As noted, a key feature of Théophile's lyric is how hyperconsciousness of the body plays a large role in writing the text. And, indeed, the present text offers a key example of what has been termed in Early Modern circles as the 'corps textuel' (Spencer 117–18). In the first strophe, Théophile alludes to the body of his subject, an old prostitute, as a textual construct:

Ceste vieille qui, des tombeaux,
Chasse les vers et les corbeaux,
Nasquit cent ans avant la guerre
Du fameux siege d'Ilion
Et avant que Deucalion
N'eust encore repeuplé la terre. (1–6)

This old hag who, from the grave,
Chases away worms and crows,

Was born one hundred years before
The famous siege of Ilion
And before Deucalion
Had yet repopulated the Earth.

What strikes the reader first in terms of Théophile's poetic self-awareness is the word 'vers' which, while most apparently referring to the worms frightened by the woman's ugliness, also clearly represents a word-play on the French term for 'verse.' Théophile's consciousness of the pun is emphasized by the fact that he employs it at the very beginning of the work as if to suggest that, despite his ribald tone, he is cognizant of the artistic circumstances surrounding the stanzas. Within Théophile's œuvre, this context is reinforced in the same line with the word 'corbeaux' eliciting a subtle reference to Théophile's 'Un corbeau devant moi croasse.' The point here is not to develop an in-depth comparison between Théophile's famous ode and this significantly lesser-known obscene poem. Rather, it shows that, even in his libertine verse, Théophile seeks to establish a certain literary authority as a poet by relating his work either to: 1) other poems he has composed, or 2) larger poetic antecedents. The first stanza contains both these micro and macro referents as Théophile again evokes the ominous presence of the crow, while the mention of the *Iliad* and Deucalion in verses four and five calls forth epic and mythological traditions. On the one hand, these last two references are significant only in that they 'jokingly' underscore the extreme age of the woman. On the other, they also suggest that the woman's hideousness, accentuated by her superannuated state, is of epic and mythological proportions. As a result, the body that inspires and writes this text is unmatched in its wretchedness.

Not content with confining the parameters of his revulsion to the epic and the mythic, Théophile hints in the next strophe that the woman's ugliness is such that she would scare even the Devil:

Un jour, ceste vilaine là
Dans un benestier distilla
Les pleurs de son œil hypocrite;
Depuis le diable, qui le vid,
Craignant de gaigner mal au vit
N'osa toucher à l'eau benite. (7–12)

One day, this wretch
Distilled in a Holy Water font
The tears of her hypocritical eye.
Upon which the Devil saw her,
And fearing a sore cock from the clap,
Dared not touch the blessed water.

The idea that a supernatural figure such as Satan could catch a venereal disease by dipping his hand in Holy Water containing the prostitute's tears reveals the magnitude of the woman's repulsiveness. At the same time, the phrase 'mal au vit' ('a sore cock from the clap') establishes the scatological nature of the poem. Already, a difference emerges in how syphilis is portrayed with respect to 'A un marquis' because Satan's vulnerability illustrates everyone's susceptibility to suffering and degeneration. Clearly, if Satan can experience this malady, then it is not something that can be cured quickly and thus allow the victim to resume his playful lifestyle. In effect, the poet himself contracts the 'mal au vit' after sleeping with the old woman, and thus becomes a 'pauvre diable' who must endure the agony of the disease. The idea of a cavorting marquis who goes from one conquest to the other is demolished. Venereal sickness is no longer to be brushed aside. Rather, its consequences have become grave, if not lethal.

Théophile is fascinated with the disease and the woman who spreads it. In the poem's third stanza Théophile seemingly revels in describing the ill woman, highlighting in particular the scatological aspects of her being:

> Ceste vesse, quand on la fout,
> Descoule de sueur par tout:
> Elle rotte, pette et [se] mouche;
> Si par fois, elle vesse aussi,
> On ne sçait lequel a vessi,
> Du cul, du nez, ou de la bouche. (13–18)

> This old fart, when you fuck her,
> Oozes sweat everywhere:
> She belches, farts, and wipes her nose;
> Should she cut a silent, stinking fart,
> One knows not from where it came,
> The ass, the nose, or mouth.

For the poet, the wizened prostitute has no essence other than her scato-sexual existence. Metonymically, he reduces her to a silent but pungent fart, or 'vesse,' in verse 13. The poet's recounting of the woman's emanations from her ass, mouth, and nose can certainly be interpreted as imitating the late sixteenth- and early seventeenth-century lyric tradition of misogynistic satire. Yet, the detail with which the narrator describes the whore's burping, flatulence, and nose-wiping manifests an obsessive quality that bespeaks mental as well as physical illness on Théophile's part. In effect, the strophe's opening verse, 'Ceste vesse, quand on la fout' ('This old fart, when you fuck her') indicates the poet's perverted desire. If the metonymy of substituting the woman for the fart is temporarily cast aside in favor of a more literal interpretation, then what the poet screws is the fart itself. As a result, what Théophile is attracted to is the fart, as well as the sweat, belching, and snot. Of course, the stanza represents a sadistic mocking of the woman by the poet. But from a lyric standpoint, this type of ridicule stands as a reaction against the tradition, to which Théophile himself adhered, of venerating the woman's body. With respect to the mouth and nose, readers need only think of the *blasons* of Marot, Scève, and Ronsard to understand the antecedents the poet seeks to undermine. The poem cannot be considered a *contreblason* because while it does insist on the woman's ugliness, it does not exclusively rail against one part of the woman's body. Because this poem essentially describes a perverse attraction, it inverts the thematic and structural elements of the *blason* rather than subverts them.[6] In this case, inversion is meant as a reversal in the use and meaning of signs. Here, disease and hideousness excite rather than repel, thus upending social, literary, and psychological norms.

Indeed, the poem should not be read merely as a crude *exercice de style* not only because the narrator is attracted to the woman's horror, but because he appropriates the woman's repugnance as his own. As Gaudiani notes, the poet confesses his own 'intimacy' (71) with the woman in the fourth stanza. Although the previous stanza suggests that the poet sleeps with the prostitute, the following strophe confirms it:

> Son foutre jaune, verd, et bleu,
> De morve, de cole et de gleu,
> Sentoit le souffre et le bytume,
> Qui descoulait sur mes couillons,
> Comme deux pestilans caillons
> Qui jaillissent d'une apostume. (19–24)
>
> Her muck yellow, green and blue,
> Made of snot, cum, and glue,
> Smells of sulfur and bitumen,
> That flow on my balls,
> Like two pestilent clots
> Gushing from an abscess.

With the reference to 'mes couillons' ('my balls'), the poet signals that his body, as well as the woman's, write the text. Clearly, he wishes that the diseased mush from the prostitute's genitals flow onto his testicles so that he too may become infected if he is not already. The narrator's likening of the odor from the whore's muck to that of sulfur and bitumen strikes the reader as a covert reference to Satan. Théophile's poet relishes the description of this sexual encounter not only for its Baroque grotesqueness but because he has achieved sexual union with a woman whom even the Devil himself has rejected. The narrator has reached the nadir of crudity by distastefully inverting poetic convention, and by personally entering into a liaison with a woman whose repugnance reaches supernatural proportions. Consequently, the poem reveals a perverse psychology at the heart of Théophile's libertine project.

An essential element of this psychology is the poet's death wish. References to Satan suggest an obsession with mortality, as well as a desire for suffering and punishment in this world, and potentially in the next. As a consequence, the sadistic enjoyment the narrator experiences from describing the woman is complemented by the masochism displayed in the poet's urge to bed the old whore and to smother himself with her filth and disease. In referring to his testicles as 'deux pestilans caillons' ('two pestilent clots'), the narrator calls attention to the deadly quality of his union with the old prostitute. The temporary ecstasy of the *petite mort*, underscored by Théophile's use of the word 'jaillir' in verse 24, translates into the permanent orgasm that is death itself. Théophile depicts the fluids of both male and female orgasm as rendering death as opposed to life. In effect, the coupling the poet depicts inverts the traditional purpose and representation of heterosexual union in that the bonding engenders not human generation but degeneration. In the fifth stanza the poet's remarks that the whore's chancres 'percent jusqu'à la matrice' ('pierce all the way to the womb') (30), meaning she is sterile. The result of the mating is not an infant, but rather two sickened, and eventually dead bodies. *Eros* converges with *Thanatos* as the poet's attraction to the aged lady of pleasure betrays in fact a fascination with death.[7]

The erotic obsession with dying does not stop with the poet. Given the narrator's description of her, the reader assumes that the tart excites herself with the knowledge of her diseased body and the damage it inflicts on others. Théophile's last two stanzas illustrate this idea:

> Son con vilain, baveux, suant
> Et plus que le retrait puant,
> Cizelé de la cicatrice
> De chaude pisse et de poullins,
> Et de mille chancres malins
> Qui percent jusqu'à la matrice.
>
> Mille morpions, rangez aux bords,
> Tous plats, pattus, et demy-morts,
> Tenoient leur general concile,
> Pour ronger l'onguent verolé
> Qui leur a quatre fois volé
> Le poil qui leur servoit d'azile. (25–36)
>
> Her ugly cunt, drooling, sweating
> And more than a stinking latrine,
> Chiseled with the scar
> Of the burnt piss and lice
> And a thousand nasty chancres
> That pierce all the way to the womb.
>
> A thousand crabs, lined up on the edges,
> All flat, with big claws, and half-dead,
> Hold their general council
> To gnaw away at the slimy clap-ridden man
> Who four times stole them
> From their bushy refuge.

The prostitute's wetness can certainly be ascribed to her desire, but the poet's description of her 'con vilain' ('ugly cunt') personifies the whore's genitalia and gives some insight into the woman's character. As is the case with Petrarch and his

Pléiade disciples, the female object of Théophile's desire is voiceless and to a large extent faceless. However, the personification of the chancres as 'malins' underscores the prostitute's ingenuity and delight in soliciting the desire of a man such as the narrator. As a result, the trollop might be nameless, but her personality is not entirely invisible. Certainly, in Théophile's mind, the woman has no function or identity other than a sexual one. But the notion that she is capable of villainy and shrewdness indicates that she sadistically enjoys delivering her services to masochists such as the poet. To suggest that the whore's body reveals parts of her psyche adds depth to the text that is her body. By describing the influence the woman's body has on his own physical and psychological being, the poet illustrates how one body writes another.

Logically, the final chapter of the bodily text is of course death. Théophile's last stanza amplifies the death imagery of the fourth stanza by portraying the 'mille morpions' ('a thousand crabs') as 'demy-morts' ('half-dead'). In addition, the poet describes the crabs as 'rangez aux bords' ('lined up on the edge') and as holding a 'general concile' ('general council'), which suggests that the prostitute's genitalia, and therefore the woman herself, have planned an attack whereby the pox will 'ronger l'onguent verolé' ('gnaw away at the clap-ridden man') that is the narrator, and assure his death. From a structural standpoint, the reference to the 'concile' echoes the poet's religious allusions to the 'benestier' ('holy water font') and to Satan at the beginning of the text. With this type of closure, the Church is again satirized, but at the same time one of its most hallowed practices is ironically respected in that he poet confesses that he has slept with the whore four times. While the poet does not become penitent in his admission, he does nonetheless make a point affirming this relationship a second time. By confirming his multiple trysts with the antiquated harlot, Théophile underscores his masochistic obsession with the woman, while engaging in a form of self-loathing. By identifying himself in the previous verse as 'l'onguent verolé,' the poet brings himself down to the same level as the whore. Contrary to many poets in the Early Modern satirical tradition who berate women as a means of vaunting themselves, Théophile's air of superiority – which is clearly present at the beginning of the poem – disappears by the end. His existence, like that of the prostitute, is reduced to scatological functions. Now *he* will have to deal with the burnt piss, the chancres, and the viscous discharge. While the poet may indeed pathologically wish this lot upon himself, his manner of self-representation is unflattering by any objective standard. The syphilis he has contracted will transport him from the 'azile' that is the woman's sex in the last verse, to that of the sanitarium. The text thus represents a self-portrait as much as it does a depiction of the weather-beaten prostitute. Given that the narrator concludes the poem with an image of his decaying body, the work could just as easily been entitled 'Stances contre un verolé' ('Stanzas against a clap-ridden man') or even 'Stances contre moy-mesme' ('Stanzas against myself'). Théophile's originality lies in turning a satirical diatribe against a tart into an invective against himself. Not only does he recount personal experience, but also personal destruction in a genre where verbal annihilation of the other is the norm.

Dysfunction and Despair: A World without either Sex or God

A remarkable feature of Théophile's cabaret poetry – and one which shows that it should not be dismissed as so much obscene verse – is that the reader can detect significant shifts in tone and theme within the genre. One poem which stands out in tracing these variations is 'On m'a dit que ma sœur chevauche' ('They tell me my sister's a whore') in which Théophile's representations of the scatological, the philosophical, and the poetic converge to form a world-view that both coincides with, and deviates from, the perspectives outlined in the two previous poems. The work's second quatrain quickly establishes the tenor of the work:

> Que nos jours ont un mauvais sort;
> Que ma planette est mal logée;
> Que la fortune est enragée
> De me persécuter si fort! (5–8).
>
> That our days meet a sorry end;
> That my planet is misaligned;
> That fortune is so enraged
> To persecute me so harshly.

What this stanza typifies is the poet's anger, which he expresses on both a collective and personal level. The reference to 'nos jours' ('our days') as having a 'mauvais sort' ('sorry end'), as well as the implication that the poet himself is 'persecuted' by 'fortune' indicates that Théophile's narrator views his individual misfortune as representative of humanity's as a whole. With respect to 'A un marquis' and the 'Stances contre une vieille,' 'Ma sœur' reveals none of Théophile's playfulness nor excitement, no matter how perverse. Generically, Théophile labels the poem a 'plainte' ('complaint'). And when he exclaims in verse 9, 'Je me fasche, et me plains de tout,' ('I get angry, and complain of everything') and in verse 11, 'Ventrebleu! le destin me fout' ('Damn it! Fate fucks me over'), it becomes clear that the poet's capacity for enjoyment and/or subjectivity has been significantly diminished. The world is now screwing the poet, not the other way around. Without question, such a tone reinforces the idea of self-destruction seen in the 'Stances contre une vieille.' And while such an assertion is true to a large extent, the poet's cry of 'Tout me desplaît' ('Everything displeases me') in verse 10 suggests that now there is no pleasure in misery. As in the two other poems, Théophile focuses on venereal disease, but varies its depiction. Now, not only can syphilis not be cured, as it is in 'A un marquis,' it has become so severe that it has rendered the poet all but inactive from a sexual point of view. In the 'Stances contre une vieille' the idea of death from a sexual disease is certainly present, but it does not stop the poet from engaging in aberrant erotic delights. Here, however, the poet's malady is so advanced that he renounces sexual activity all together: 'Foutre des culs et des garçons! / Maugrebieu des cons et des garces!' ('Fuck all assholes and boys! / To hell with cunts and bitches!') (23–4). Of course, the use of the term 'foutre' is ironic in this case because it expresses the poet's desire to be rid of all sexual orifices rather than penetrate them. As a result of Théophile's sexual dispossession, he appropriates the Baroque motif of the 'corps désanimé' where the virtually 'lifeless body' comes to represent a kind of 'cadavre parlant' ('talking cadaver').[8] Théophile's narrator is much closer to death in this poem as opposed to the last, infusing the scatological context of the poem with a larger dose of *Thanatos* than before. In effect, the general desexualization of the poet brought about by the disease is such that the scatological is no longer sexual, as it was in the 'Vieille,' because the presence of death is so dominant. The lament's fourth and fifth stanzas highlight the narrator's desperation:

> Je pisse du verre et du feu;
> Je ne crache que de la colle;
> Je n'ay presque plus un cheveu:
> J'ay la peste! j'ay la verolle!
>
> La gravelle me tient aux reins;
> Je ne trouve plus qui je foute,
> Et la saincte ampoule de Rheims
> Tariroit plutost que ma goutte. (13–20)
>
> I piss glass and fire;
> I spit only grunge;
> My hair is almost gone:
> I have the plague! I have the clap!
>
> Stones have a hold on my kidneys;
> I can't find anyone to fuck,
> And the holy ampulla of Reims
> Will dry up faster than my gout.

With respect to the language of the other poems studied, the reader notes that direct mention of the genitalia – either his or someone else's – is missing. With one exception late in the poem, gone are the references to the 'vit,' the 'con,' and the 'cul,' as well as the excitement they elicit. Instead, the poet alludes to the deterioration of the genital area caused by the disease. Théophile describes similar afflictions in the 'Vieille,' but the ravages of the illness do not hinder his activity. If anything, they enhance it by increasing his exhilaration. The very nature of the disease is more extreme in the current poem because here the poet is pissing 'glass and fire,' as opposed to dealing merely with the 'pisse chaude' ('burnt piss') alluded to the 'Marquis' and the 'Vieille'. The internal contradiction that is the image of pissing fire underscores Théophile's desperation. Within the context of the lament, this represents the narrator's hopelessness because an act that normally snuffs a flame, i.e., that of passing water, only serves to intensify the burn. Evacuation, rather than relieving the torment, only exacerbates it. Likewise, the allusion to 'pissing glass' advances the idea that, though the poet's corporeal deterioration may start with the genital area, it is not limited to that region. The image foreshadows the reference to the 'gravelle...aux reins,' otherwise known as a kidney stone. Scatological allusions signal that the disease has spread to the narrator's vital organs, and will ultimately consume him.

The theme of waste associated with Théophile's scatological references expresses itself in the realm of sex as well. A declaration such as 'Je ne crache que de la colle' ('I spit only grunge') further develops the idea of the narrator's de-sexualization. The word 'cracher,' a vulgar term for 'éjaculer,' suggests that the poet's orgasms are devoid of any agreeable sensations. To come is to expel, nothing more. The discharge is a glutinous substance only referred to as 'colle.' With language of this nature, it becomes difficult to distinguish semen from phlegm, as the body makes a futile attempt to rid itself of infection by any means possible. Sexual dysfunction corresponds to mental dysfunction in verse 18. The poet's statement, 'Je ne trouve plus qui je foute' ('I can't find anyone to fuck') is rich in meaning. On the one hand, it suggests that because of the narrator's physical ailments, no one will have sex with him. On the other, it implies that Théophile's mental faculties have declined to the point where he is incapable of finding his bed partner, perhaps even in the same room.

If read alongside 'A un marquis,' an additional explanation for the narrator's impotence emerges in the lament: God's impotence. Throughout the poem, Théophile portrays a God whose indifference to human fate borders on the malignant:

> Je n'attends secours d'aucun lieu;
> Mon malheur est insatiable;
> Les hommes m'envoyent à Dieu,
> Qui m'assiste autant que le diable. (29–32)
>
> I expect help from no quarter;
> My misfortune is insatiable;
> Men send me to God,
> Who helps me as much as the Devil.

Here, God is no better that Satan in terms of his response to human suffering. This portrait stands in marked contrast to the libertine God of the 'Marquis' who encourages humanity to follow its natural inclination toward sex. In the present poem, God stifles human sexuality, preventing erotic interaction between his Creatures:

> Dans le commun lict des humains
> Un Dieu veut que toujours on dorme.
> Ces beaux vits et ces belles mains
> N'ont plus ni mouvement ni forme. (37–40)
>
> In the common bed of humanity
> A God wishes that we always sleep.
> That these beautiful cocks and hands
> Have neither movement nor form.

The stanza's tone is wistful in that the sensual appendages of the human body are rendered useless by a cold deity who puts humankind in the same bed only to make it sleep. Clearly, the image of the motionless, sexless lot that is humanity suggests that humankind is doomed, if not dead, from the start. By negating sex, God cuts off the source of all life, leaving his progeny to rot in the manner of Théophile's poet. Divine nature thus runs contrary to human nature, as God, who is in essence asexual, destroys humankind whose very existence is defined by its sexuality.

Théophile's mention of God in both the 'Stances contre une vieille' and in the present poem raises the question of how this portrayal compares to that found in the religious lyric of the Baroque era. On one level, no comparison at all exists because Théophile's references to God are satirical at best and – as his trial and death sentence no doubt indicate – blasphemous at worst. As a result, there is little basis upon which to read Théophile's cabaret poetry within the context of the Early Modern devotional lyric. Yet, issues of illness, sexuality, and the human condition in general show that the philosophical dimension of Théophile's lyric is readily perceptible. And as Gaudiani has shown quite convincingly, this particular lament by Théophile imitates the format of composition, analysis, and meditation found in Baroque religious poetry (50–51).[9] From a comparative standpoint, the accent on decomposition links Théophile's work to the penitential lyric and to the *memento mori* traditions of the late Renaissance and early seventeenth century.[10] One trait that perhaps rather surprisingly links Théophile's libertine verse with the religious lyric is the use of the scatological as it relates to sickness, dying, and God's role in human existence. As can be imagined, devotional poets, who often sought Church approval for their verse, were limited in the kind of language they could deploy to describe human physical and mental degeneration. Nonetheless, poets such as Philippe Desportes (1546–1606) use terms such as 'crasse' and 'ordure' to describe their internal and external deterioration:

> Mais quand je me regarde au miroir de ta loy,
> Que dedans et dehors transformé je me voy,
> Que je trouve en mon ame et de crasse et d'ordure!
> Que mes sens corrompus sont devenus infectes.... (quoted in Cave 121)

> But when I see myself in the mirror of thy law,
> That inside and out transformed I see myself,
> That I find in my soul filth and mire!
> That my corrupt senses have become diseased.

Without question, the same terminology could be used to describe Théophile's physical and mental states. Théophile's language is more explicit simply because of the nature of the cabaret genre. However, given the devotional tenor of Desportes's lyric, the appearance of words such as 'crasse' and 'ordure' in a religious poem would have produced effects of shock and aversion in the devotional reader, much in the manner of Théophile's libertine œuvre. In both cases, the poet's approach is similar: to provoke, and in some cases offend, the reader into re-orienting his or her thinking with respect to poetic form and the manner in which it expresses the human condition.

Desportes's somewhat muted scatological vocabulary notwithstanding, Auvray makes no effort to mask the part scatology plays in the Crucifixion, where he describes the dying Christ as 'de toutes parts pissait des flots de sang' ('[having] flows of blood pissing from everywhere') (quoted in Debaisieux 136). The idea of Christ 'pissing blood' from all parts of his body renders the idea of Christ's sacrifice more comprehensible by appealing to the basest of human functions. In this case, the incorporation of the grotesque tradition, or that which seeks to evoke horror or disgust, has a didactic function in that Christ, in one of his most divine moments, commits one of the most human of acts. As Debaisieux suggests, the purpose of such a portrait is to heighten the reader's sense of 'participation sensorielle et affective' ('sensorial and affective participation') (135). What the scatological does, in both the religious lyric and the cabaret lyric, is enable the reader to identify with the poetic subject – whether it be Christ in Desportes and Auvray, or with Théophile himself in the latter's poetry. Of course, the devout public has little idea of what it is like to die on a cross, just as presumably few of Théophile's readers have had firsthand experience of syphilis. However, in both instances, the scatological can be traced to the suffering and decay which are an intractable part of the human condition. While alluding to humanity's future encounter with death, the scatological also suggests humankind's present vulnerability, where man finds himself particularly helpless in his inability to control the body. Christ's devout body, as well as Théophile's libertine body, are equally exposed to agony and deterioration. The very fact that the body produces waste indicates that humanity is incapable of mastering the physical dimension of its being. From a philosophical standpoint, the inability to master the body becomes indicative of the incapacity to master the self. Montaigne argues for the submission of the body to the mind – a sentiment later echoed by Descartes.[11] In the neo-Classical period, the notion of the controlled body forms an integral part in the development of the self-possessed individual and state. During the Baroque era, however, the body eludes this type of dominance, showing itself easily weakened by humanity's fragile character. In effect, this section of Théophile's lyric shows an anti-Humanistic strain by underscoring the distinct limits of the human body and mind. Human capacity is thus decidedly curtailed by physical and metaphysical constraints.

For the devotional reader, humanity's refuge in the face of physical and spiritual misery is, of course, God and his salutary promise. For Catholic poets such as Desportes and Auvray, the scatological nature of man's existence is indicative of original sin and the subsequent need for redemption. If humanity is able to recognize itself as filth, then it can purge itself of sin and hope to attain grace. In discussing the work of the Counter-Reformation theologian Fra Giacomo Affinati d'Acuto, Piero Camporesi speaks of 'Mortification both internal and external of the self [as] the right key to open the gates of the Kingdom of Heaven' (73).[12] Similarly, early seventeenth-century Catholic theology and poetry claimed that 'decay and dissolution will be overcome' by faith in the resurrected body of

Christ and in the social body of the Church (Coakley 128).[13] In Théophile's anti-theistic universe, bodily suffering in the hope of eventual divine blessing is clearly not an issue. However, in the absence of salvation, could some form of purgation and/or consolation be possible? At first glance, seemingly not. Sexual relief offers no hope, and even poetry itself appears to bring little solace. As the narrator states toward the end of the poem, in the ninth stanza:

> Vaines ombres de l'Achéron,
> Larves, démons, rivages sombres,
> Pétrone, Arétin, Maugiron,
> A grand'peine êtes-vous des ombres! (33–6)
>
> Vain shadows of the Acheron,
> Worms, demons, somber shores,
> Petronius, Aretino, Maugiron,
> At best you are only shadows!

Satirical and bawdy predecessors, such as Petronius, Aretino, and Maugiron are of no help to Théophile.[14] In fact, by associating these authors with Acheron and other images of hell, Théophile suggests that not only have his lyric mentors failed him, they have perhaps even played a role in leading him to the desperation in which he now finds himself. The three poets must share in the blame by setting an example in which life and art converge in such a way as to bring about the poet's downfall. In 'On m'a dit que ma sœur chevauche' at least, the act of writing does not help in purging the narrator of his anguish. Neither sex, God, nor poetry provides the opportunity for catharsis. If anything, it seems as though painful sex, an uncaring deity, and the misleading poetic examples that have preceded Théophile only serve to increase his torment.

Conclusion: Sarcasm as Temporary Salvation

Nonetheless, 'Ma sœur,' despite the unexpectedly articulate and nuanced manner in which it expresses the fate of humankind, should not be seen as Théophile's definitive pronouncement on the subjects of sex, illness, and God. Much of his cabaret poetry reveals a sarcastic tone that calls to mind not so much hope or solace as it does the idea that human wit can survive even in the most exasperating of circumstances. One such example is a sonnet that takes the form of a sardonic lament:

> Phylis, tout est foutu, je meurs de la verolle;
> Elle exerce sur moy sa derniere rigueur:
> Mon vit baisse la teste et n'a point de vigueur;
> Un ulcère puant a gasté ma parole.
>
> J'ay sué trente jours, j'ay vomy de la colle;

Jamais de si grands maux n'eurent tant de longueur;
L'esprit le plus constant fust mort à ma langueur,
Et mon affliction n'a rien qui la console.

Mes amis plus secrets ne m'osent approcher;
Moy-mesme, en cest estat, je ne m'ose toucher.
Phylis, le mal me vient de vous avoir foutue!

Mon Dieu! je me repens d'avoir si mal vescu,
Et si vostre courroux à ce coup ne me tue,
Je fais vœu desormais de ne foutre qu'en cu!

Phylis, all is fucked, I'm dying of the clap;
It imposes on me its final rule:
My cock dips its head and has no vigor;
A stinking ulcer has ruined my speech.

I've sweated for thirty days, I've vomited muck;
Never have such great ills lasted so long;
The firmest mind would be overcome by my langor,
And my affliction has nothing to console it.

My most intimate friends dare not approach me;
Even myself in this state I dare not touch.
Phylis, evil has come to me from fucking you!

My God! I repent for such foul living,
And if your anger at this time does not kill me,
I hereby swear to fuck only in the ass!

Images of illness and death dominate the beginning of the sonnet, with Théophile mocking the now-familiar references to impotence, vomiting, and general agony. Not only is he incapable of sex with another, he is, according to verse 10, powerless even to masturbate. The narrator is alone, abandoned by his friends and beyond consolation (8). From these perspectives, the poem clearly resembles the 'Stances contre une vieille' and 'On m'a dit que ma sœur chevauche' in unveiling the desperation Théophile experiences during his malady. And as in other texts, scatological and/or repulsive allusions such as 'Mon vit baisse la teste' ('My cock dips its head') (3), and 'j'ay vomy de la colle' ('I've vomited muck') (5), serve to reinforce the link between human waste and death. Indeed, Théophile continues to draw the parallel between sexuality and death when he states in verse 11, 'Phylis, le mal me vient de vous avoir foutue!' ('Phylis, evil has come to me from fucking you!'). The apostrophe to Phylis, also made in the sonnet's first verse, is significant in that this is the name of the beloved in many of Théophile's 'stances' and 'odes,' of which one of the most read is 'Le Matin.' By attributing responsibility for his sickness to Phylis – clearly a word-play on 'Si-Phylis' – Théophile not only repudiates her and the disease, but the poetry of which they are the object. 'Si/Phylis' becomes a scatological referent in that it is the cause of the narrator's ulcers (4), sweat, and phlegm-laden vomit (5). Within this logic, the poet holds 'Si/Phylis' accountable for his dementia as well, stating, 'L'esprit le plus constant fust mort à ma langueur' ('The firmest mind would be overcome by my languor') (7). The inspiration for Théophile's earlier work has now turned into a kind of anti-Muse, destroying the poet's creativity as well as his health.

However, the final tercet brings about a significant change in the sonnet's tone. The idea of making a 'bargain' with God to have only anal intercourse if the Almighty spares his life removes any trace of seriousness the poem might contain.[15] The tenor of 'A un marquis' is considerably light-hearted, but a decidedly negative shift occurs in the 'Stances contre une vieille' and in 'On m'a dit que ma sœur chevauche' because the poet's irony disappears behind personal distress and existential angst. In the current poem, the mock prayer satirizes the devotional lyric of the Baroque era, while showing that, despite his suffering, Théophile has retained his verbal bite. Of course, Théophile will succumb to his illness. But for a moment at least, he preserves the caustic attitude at the heart of many of his satirical works. The steep modulations in tone throughout the poem and in Théophile's œuvre lend a significantly protean quality to his lyric. To a certain extent, Théophile's cabaret lyric can also be thought of as subverting certain aspects of the religious Baroque in that sensuality leads not to death and redemptive ecstasy but to pain and death as ends in themselves. Spirituality, if it is to be considered at all, should be looked upon with derision or incredulity. As far as the relationship between the Baroque and the scatological is concerned, attention to human refuse allows the poet to concentrate on the reality of the body in order to subvert the illusions of the soul. Baroque detail and excess stay at a human level, putting man in his place not with respect to God, but with respect to himself. Any catharsis that might be achieved through suffering, poetry, and/or humor, begins and ends with the poet, underscoring that only in this world can humanity denigrate and celebrate its foibles.

Notes

[1] I use the terms 'cabaret' and 'libertine' interchangeably. Both refer to the scabrous verse Théophile wrote as part of the bawdy tavern culture in Paris during the early seventeenth century. See Gaudiani, *The Cabaret Poetry of Théophile de Viau* 15–22. All quotes from Théophile are taken from this edition; English translations of Théophile are mine.

[2] In foregrounding his thesis on the body and changing social order, Greenberg speaks of 'the Rabelaisian view of the universe as an ever-evolving living entity, constantly in process, copulating, birthing, and dying without beginning or end' (62). He also cites Montaigne's *Du Repentir* (III.2): 'Le monde n'est qu'une branloire perenne. Toutes choses y branlent sans cesse: la terre, les roches du Caucase, les pyramides d'Aegypte, et du branle public et du leur. La constance mesme n'est autre chose qu'un branle plus languissant' (62).

[3] Deterioration of the body as it relates to the Baroque aesthetic has been the subject of two recent articles, see Safty and Souiller.

[4] The appendix of Guido Saba's critical edition of Théophile's *Œuvres poétiques* contains many of Théophile's licentious poems, but gives few critical remarks. For a history of the general reception of Théophile's lyric, see Saba's *Fortunes et infortunes de Théophile de Viau.* Joan DeJean's *The Reinvention of Obscenity* deals with Théophile's ribald lyric, but came to my attention after this essay had been prepared for publication.

[5] Debaisieux's *texte de base* is Auvray's *La Pourmenade de l'Ame dévote en Calvaire, Accompagnant son Sauveur depuis les ruës de Jerusalem, jusqu'au Tombeau*, Rouen, 1634.

[6] For a more precise definition of the *blason* and the *contreblason*, see Persels, 'Venereal Verse.'

[7] *Thanatos* is, of course, the Greek figure representing death. Western thought has often opposed *Thanatos* and *Eros*. However, Théophile views them as intertwined.

[8] I take the term 'corps désanimé' from Soare, 'Cadavres exquis du théâtre baroque.'

[9] Gaudiani follows the tripartite structure that forms the core of Louis Martz's argument in *The Poetry of Meditation*.

[10] For more on these devout motifs, see Chapter 5 of Terence Cave's *Devotional Poetry in France*.

[11] See Dandrey 42–3, where he discusses Montaigne's *Apologie de Raimond Sebond* and *De trois commerces*. The theme of mental domination of the body is found throughout Descartes's œuvre. However, for a precise explanation of the mechanics of the subordination of the body to the mind, see the fourth and fifth parts of the *Discours de la Méthode*.

[12] D'Acuto's text is entitled, *Il Monte Santo della Tribulatione* (*The Holy Mount of Tribulation*), published in 1602.

[13] Coakley cites especially C.W. Bynum's 'Bodily Miracles and the Resurrection of the Body in the High Middle Ages.'

[14] Greenberg notes that Aretino's *Ragionamenti*, a sexual coming-of-age novel published in 1534, was known throughout Early Modern Europe and 'decisively influenced ... the modern pornographic genre' in France and elsewhere (70).

[15] One is reminded of a similar vow by a syphilis-stricken narrator in Jean Molinet's (1435–1507) *Complainte d'ung Gentilhomme à sa dame*: 'Doresnavant vivray par rigle et par compas, / Au moins je ne feray jamais pour ung con pas' (69–70) ('From now on I shall live by rule and by measure,/At least I will not pass for a fucking idiot'). See *Les Faictz et Dictz de Jean Molinet* 731.

Chapter 7

Scatology as Political Protest: A 'Scandalous' Medal of Louis XIV

Jeanne Morgan Zarucchi

The commemorative medal was an important artistic genre during the reign of Louis XIV (1643–1715), as documented by the fact that one of the first Academies he created was one responsible for inventing inscriptions on medals and monuments.[1] The medal genre was ideal for linking the official *persona* of Louis with the emperors and gods of Classical antiquity. Unlike other forms of portraiture, the physical permanence of the medal conferred immortality upon its subject, while being far less expensive to produce than a statue, and reproducible in large quantity. As Louis Marin has noted, 'the medal's mystery conceals a secret that is at once a political ruse and a juridical ritual, the secret of a rational magic by which the power of state constitutes its memory in its representation and institutes it in the authority of truth' (129). This article will study how the medals of Louis XIV, an obvious form of propaganda, provided inspiration for a clever form of counter-propaganda, one work artfully concealed inside another, and thus playing a double trick upon the hapless reader.

While commemorative medals of princes and kings had been prolifically produced since the Renaissance, a distinguishing aspect of the medals of Louis XIV was the fact that engraved images of them were printed in catalogs, between 1689 and 1723, thus allowing a much wider audience to have access to them. The first catalog of the medals of Louis XIV was created in 1689, not by the Academy, but by Claude-François Menestrier, a Jesuit priest and artist who had written several important treatises on emblems.[2] The Academy immediately undertook to produce its own catalog, but the process moved slowly, and the Academy's first catalog did not appear until 1702. The Academy's second and final catalog of 1723, printed eight years after Louis's death, included 318 medals representing a chronological history of the king's life and reign.[3]

Mark Jones, former curator of the Department of Coins and Medals at the British Museum, published a monograph entitled *Medals of the Sun King*, in which he made the remarkable assertion that some of the medals illustrated in the catalogs had never been issued at all (23). This was based upon his observation that some of the engraved images in the catalogs were significantly altered from the images to be found on the same medal in the museum's actual collection, and while it was reasonable to suppose that some medals might have been reissued, it was more difficult to explain why some of the images in the catalog did not correspond to any of the Louis XIV medals in the museum, which numbered close to a thousand.

I have subsequently pursued research on the two catalogs of medals produced by the Academy of Inscriptions and Medals in 1702 and 1723, exploring a number of ways in which the truthfulness of those catalogs may be challenged. For example, the printed catalogs provided a relatively inexpensive means of refining or improving upon a medal that had already been cast. In addition, the catalogs indicate the date of the event, but not the date when the medal was actually issued, which in some cases may have been decades after the fact or not at all. The real purpose of the catalogs, therefore, was not to *inventory* a collection of art objects, but rather to *invent* a chronological history of the king's seventy-two-year reign, in which each and every year was one of military victory and personal triumph.[4]

Menestrier's earlier published catalog of the medals of Louis XIV, the *Histoire du Roy Louis le Grand* of 1689, consists of engraved images of the obverse and reverse of each medal, with six images (i.e., three medals) represented on each recto page. On the opposite verso, there is text which identifies the 'First,' 'Second' and 'Third' medal, citing the medal's Latin inscriptions and providing not only a French translation of the Latin, but also a verbal description of the image, detailing the allegorical figures represented, and the actions in which they are engaged. For many medals, Menestrier's text went far beyond a simple description, to include an admiring and highly embellished interpretation. For example, a medal commemorating the 'reestablishment of military discipline' in 1665 shows the king reviewing his troops; the text says that 'the King, since the year 1665, trained his Troops by frequent Reviews in time of Peace, in order to make good use of them in time of War with these Conquests which have astonished all of Europe' (12).

In another example of a medal from Menestrier's work, there is an obvious anachronism which illustrates the fictional aspect of all of the medals catalogs. The 1638 medal honors the king's childhood, and the vow by Queen Anne of Austria to build a church if she should bear a child (*ANNA DEI GRATIA FRANCIAE ET NAVARRAE REGINA ... OB GRATIAM DIVINI, DESIDERATI, REGII ET SECUNDI PARTUS*) (5). The medal's obverse lists the king's birthdate of 5 September, 1638. The image on the obverse, however, is of the Church of Val de Grâce, for which the young king laid the first stone in 1645. François Mansart's masterwork was not completed until 1669, three years after the Queen's death (*Le guide du patrimoine* 545–48). While it is true that the medal could have been designed from the architect's plans, the depiction of the church assumes its actual completion at the time the medal was cast. It is also interesting to note that this medal does not appear in either of the two catalogs of the Academy of Inscriptions and Medals, perhaps because it celebrates an action of the Queen, not an action of Louis himself.

During the course of my research at the British Museum, I encountered a copy of Menestrier's work that was apparently a pirate edition, produced probably in the Netherlands.[5] A close examination of the title page (Figure 7.1) showed that

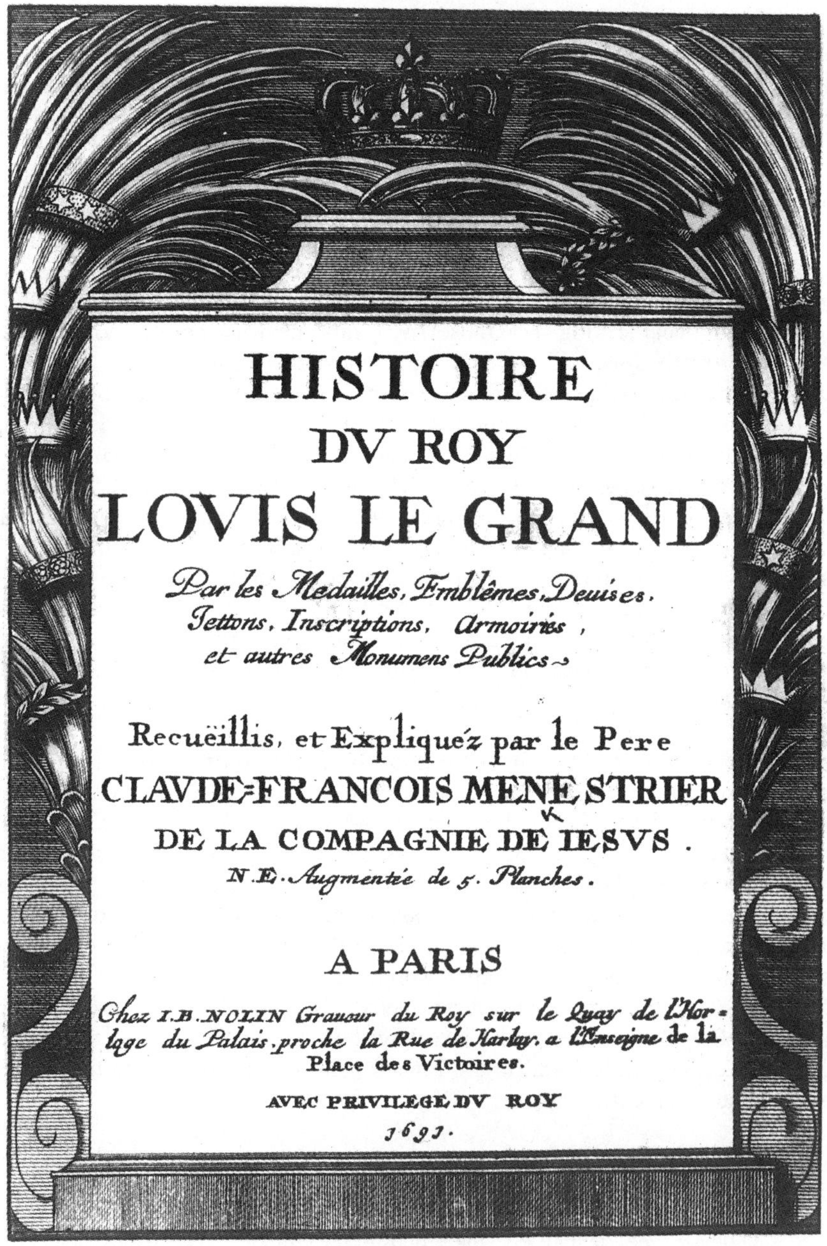
HISTOIRE
DV ROY
LOVIS LE GRAND
Par les Medailles, Emblêmes, Deuises, Jettons, Inscriptions, Armoiries, et autres Monumens Publics.
Recueïllis, et Expliquéz par le Pere
CLAVDE-FRANCOIS MENESTRIER
DE LA COMPAGNIE DE IESVS.
N.E. Augmentée de 5. Planches.
A PARIS
Chez I.B. NOLIN Graueur du Roy sur le Quay de l'Horloge du Palais, proche la Rue de Harlay, a l'Enseigne de la Place des Victoires.
AVEC PRIVILEGE DV ROY
1691.

Figure 7.1 Title page of of false edition of Menestrier's *Histoire du Roy Louis le Grand*, with added line 'N.E. [Nouvelle Edition] Augmentée de 5 Planches' ('New Edition with Five Additional Plates'). British Library Printed Books 1488.dd.6, by permission of the British Library.

the engraving plate from which it was printed was hand-lettered, probably imitating the printed title page of the original, but incorporating a few changes. Whereas the book's printed royal privilege states that it was granted in 1689, the title page says 1691. The title page also states, below Menestrier's name, that the book is 'Augmented with Five Pages.'

The five added pages are 38 to 43. The numbering of these pages follows consecutively from the preceding ones, so it is probable that five real pages were removed in order to make space for the false ones, to prevent the reader from suspecting that the book has been altered. The choice of these page numbers may have no particular significance, but 1638 was the date of Louis's birth and 1643 the date when he became king. For those five years, he was merely mortal, not yet elevated to royal status, and perhaps the propagandist wished to suggest that the 'truth' about Louis could be found between these two significant numbers.

The first false sheet captures the reader's attention with large printed lettering, which reads as follows: 'Les cinq PLANCHES des MEDAILLES qui suivent, ne sont pas moins curieuses à l'Histoire de Louis le Grand, que les précédentes: mais le Pére Menestrier a eu ses veuës pour ne les avoir pas inserées dans cét Ouvrage' ('The five plates of Medals which follow, are no less curious for the History of Louis the Great, than the preceding ones: but Father Menestrier had his reasons for not including them in this work') (38). The subsequent pages present engravings of fourteen apocryphal medals, in the same format used by Menestrier, including imaginary legends in Latin, with French translations and a paragraph describing the event.

All of the medals refer to events taking place between 1685 and 1690, from the Revocation of the Edict of Nantes to the Battle of Beveland, a group of former islands in the Netherlands now joined to the mainland. The images are closely imitative of the others in the volume, and only a close inspection reveals such shocking details as a dog feasting on a corpse. A medal of the religious persecution of the inhabitants of Vaud, a city in the Piedmont valley that is now in Switzerland, shows a kneeling female captive, not unlike other images representing defeated cities; but in this image, she is being burned at the stake, while a figure in vaguely ecclesiastical dress waves a misshapen wooden cross.

The text accompanying these medals is unambiguously propagandistic against the French. For the medal just described, the text translates the Latin legends as 'To the Vaudois and to the French, who are brothers in faith,' and then describes the reverse image as

> A Woman on a Pyre, and above the name Jehovah in Hebrew letters surrounded by rays of the sun, with these words, DOMINUS LIBERABIT, The Lord will deliver us. The Persecution for the sake of religion against the Vaudois, in the valleys of the Piedmont, by the Troops of the king of France, is the subject of this Medal, in which one sees a city in flames, which indicates the destruction of the valleys, and a battleship on the sea.[6]

The text of the last apocryphal medal lauds the heroism of the Dutch, despite their apparent defeat in the Battle of Beveland on 20 July, 1690:

> In this battle, there were only 22 Dutch ships, fighting against 82 French vessels, because the English did not fight except for the late Duke of Grafton and one other English captain, because of the cowardice and secret collusion of Mylord Torrington, Admiral of the Fleet, who prevented the English from fighting. ... It is certain that the bravery of the Dutch in this battle surpasses anything which has ever occurred, because they forced 82 French vessels to abandon 17 ships that were completely unmasted. This will seem to Posterity to be a Marvel that will be hard to believe.

The medal I have selected to analyze in detail for this article is the only one of these apocryphal medals which is scatological, and the text makes it very explicit that this is not for humorous effect, but rather to convey the profound revulsion inspired by the king's actions. The medal appears on a page in which six circles are drawn, but the middle circles are blank, so that only two 'medals' are represented (top row and bottom row). A description of the page follows here, with a complete citation of the text printed to the left of each image (Figures 7.2 and 7.3).

First Medal (detail, figure 7.4):
[Obverse, author's description]
Three figures are visible: at left, a man wearing a pontifical vestments kneels on the ground, holding a long-handled chamber pot in his right hand. His head is facing toward the central figure of Louis XIV, who is crouching over the pot. Louis' coat parts to reveal the cleft of his buttocks, from which he emits a stream of excrement into the pot. In his right hand, Louis holds the handle of a sword which points towards the right, crossing at knee-level in front of a standing male figure in Turkish costume, with pointed, curved shoes and a turban. The male figure holds in his left hand a pot, into which Louis vomits. On the ground, to the far right of the image, is an unidentified orb. Latin inscriptions appear in the legend and exergue. Author's translation of French text:

> The king gives up his Claims to the Pope and to the Algerians. Even the Gods cannot resist Necessity: Louis XIV, giving up to Pope Alexander VIII the Freedom of the Towns and City of Avignon, and even asking the Algerians for Peace. 1689. This Medal is scandalous. One sees on one side the king, who is vomiting into a Pot held by an Algerian, and on the other side the Pope, who is holding a syringe in one hand, and with the other presents a Basin to the rear end of His Majesty.

[Reverse]
On a small grassy mound with rocks, a ball decorated with three fleurs-de-lis is torn open by an explosive burst (shown by lines and clouds of smoke). Translation of French text:

> The huge Bomb which the king had had carried to Algiers, but which scandalously

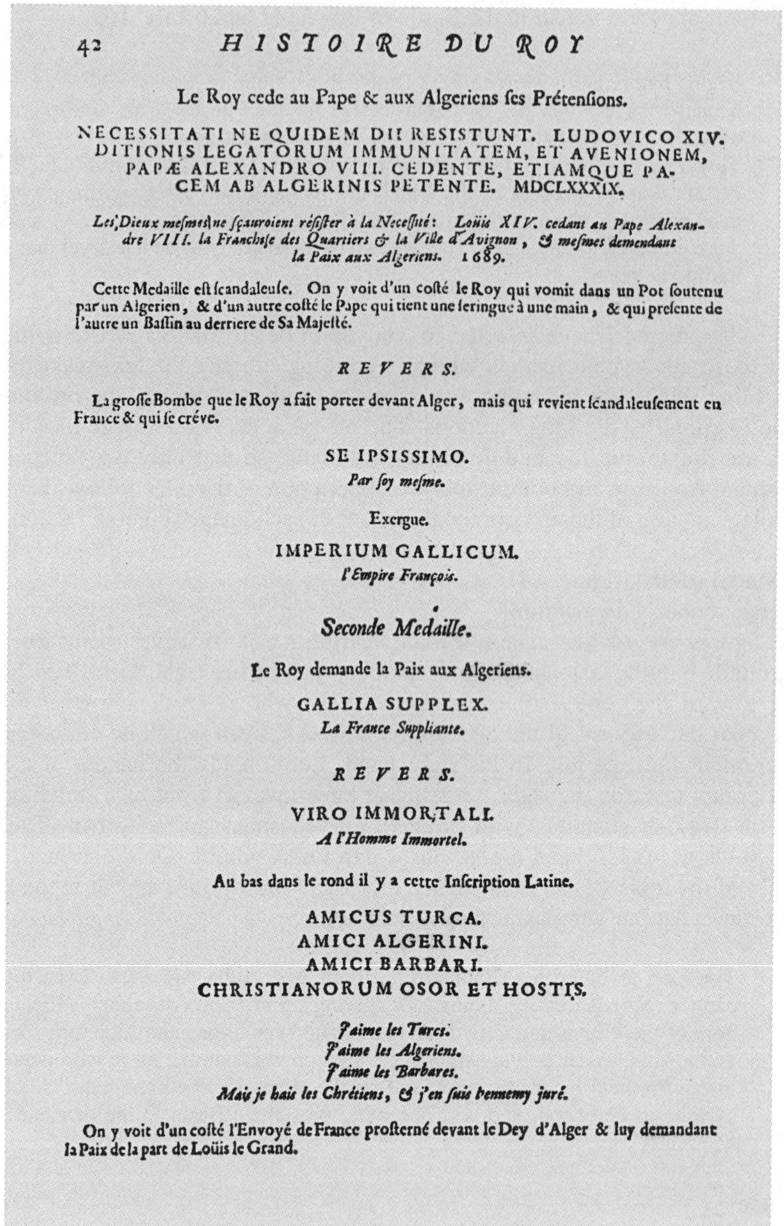

42 HISTOIRE DU ROY

Le Roy cede au Pape & aux Algeriens ses Prétensions.

NECESSITATI NE QUIDEM DII RESISTUNT. LUDOVICO XIV. DITIONIS LEGATORUM IMMUNITATEM, ET AVENIONEM, PAPÆ ALEXANDRO VIII. CEDENTE, ETIAMQUE PACEM AB ALGERINIS PETENTE. MDCLXXXIX.

Les Dieux mesmes ne sçauroient résister à la Necessité: Loüis XIV. cedant au Pape Alexandre VIII. la Franchise des Quartiers & la Ville d'Avignon, & mesmes demendant la Paix aux Algeriens. 1689.

Cette Medaille est scandaleuse. On y voit d'un costé le Roy qui vomit dans un Pot soutenu par un Algerien, & d'un autre costé le Pape qui tient une seringue à une main, & qui presente de l'autre un Bassin au derriere de Sa Majesté.

REVERS.

La grosse Bombe que le Roy a fait porter devant Alger, mais qui revient scandaleusement en France & qui se créve.

SE IPSISSIMO.

Par soy mesme.

Exergue.

IMPERIUM GALLICUM.

l'Empire François.

Seconde Medaille.

Le Roy demande la Paix aux Algeriens.

GALLIA SUPPLEX.

La France Suppliante.

REVERS.

VIRO IMMORTALI.

A l'Homme Immortel.

Au bas dans le rond il y a cette Inscription Latine.

AMICUS TURCA.
AMICI ALGERINI.
AMICI BARBARI.
CHRISTIANORUM OSOR ET HOSTIS.

J'aime les Turcs.
J'aime les Algeriens.
J'aime les Barbares.
Mais je hais les Chrétiens, & j'en suis l'ennemy juré.

On y voit d'un costé l'Envoyé de France prosterné devant le Dey d'Alger & luy demandant la Paix de la part de Loüis le Grand.

Figure 7.2 Page 42 of false edition of Menestrier's *Histoire du Roy Louis le Grand*, describing two false medals, obverse and reverse. British Library Printed Books 1488.dd.6, by permission of the British Library.

Figure 7.3 Page 42 (bis) of false edition of Menestrier's *Histoire du Roy Louis le Grand*, depicting two false medals, obverse and reverse, described in Figure 7.2. British Library Printed Books 1488.dd.6, by permission of the British Library.

Figure 7.4 Detail of the ‘scandalous’ medal of Louis XIV, Pope Alexander VIII, and the chamber pot. From false edition of Menestrier’s *Histoire du Roy Louis le Grand*. British Library Printed Books 1488.dd.6, by permission of the British

> returns to France and blows itself up. *SE IPSISSIMO*, By itself. Exergue: *IMPERIUM GALLICUM*, the French Empire.

Second Medal:
[Obverse]
There are four figures in the scene: from left, only a head, bent over in close proximity to the buttocks of the second figure, who kneels in abject prostration, arms extended forward and head bowed to the ground. At center right, a standing figure in Turkish costume holds a scimitar over the kneeler's head, and at far right, another figure in similar dress is seated, contemplating the scene with what seems to be a smile on his face. On the ground, one sees a cross, a sword, and bags spilling open to reveal coins. Translation of French text:

> The king asks the Algerians for Peace. *GALLIA SUPPLEX*, France Pleading.

[Reverse]
Hand-lettered Latin words, no image: *AMICUS TURCA. AMICI ALGERINI. AMICI BARBARI.CHRISTIANORUM OSOR ET HOSTIS*. Translation of French text:

> *VIRO IMMORTALI*, To the Immortal Man. Below, in the circle, there is this Latin inscription: *AMICUS TURCA. AMICI ALGERINI. AMICI BARBARI. CHRISTIANORUM OSOR ET HOSTIS*. I love the Turks. I love the Algerians. I love the Barbarians. But I hate the Christians, and I am their sworn enemy. One sees on one side the Envoy of France, prostrate before the Dey of Algiers and asking him for Peace on behalf of Louis the Great.

The posture of the French envoy in the second medal is revealing, as his rear end is higher than his head in a demeaning pose of submission. Furthermore, the juxtaposition of the two medals on the same page makes it clear that Louis's actions have shamed all of Christendom, uniting both Catholics and Protestants in outrage against his capitulation to the true heretics.

What historical events may have prompted this virulent criticism? There is relatively limited documentation of the events relating to French military actions in Africa, compared to the hundreds of pages devoted by historians to the minutiae of France's relations with its European neighbors. It is recorded that French naval forces bombarded Algiers in 1683 and again in July 1688, which explains the medal's image of an exploding bomb, as a metaphor for the self-destruction of the French empire. The bombardment of Algiers was intended to discourage acts of piracy which interfered with French commerce.

The connection between the Pope and these events is ambiguous. Emperor Leopold I of Austria was waging war against the Turks, and Pope Innocent XI had asked Louis to join a Christian alliance against the Turkish empire, an appeal which Louis largely ignored. The medal refers, however, to a different pope, Alexander VIII, who succeeded Innocent XI in August 1689. One month later, on 15 September 1689, French troops entered the city of Avignon, and as declared by the

Parliament of Provence, the city was officially 'reintegrated into the realm.' This was the action to which the text referred, as 'Louis XIV, giving up to Pope Alexander VIII the Freedom of the city of Avignon.' The repression of religious freedoms in the city which was the historic seat of rival popes was a symbolic and political turning point in Louis's war against the Protestants.

It is therefore possible that this medal was inspired by outrage over the military action against Avignon, and the Algerian connection is exploited in order to present Louis as having committed a dual capitulation, both to the Catholic Pope and to the heretic Turks. This concept of capitulation may explain the positioning of the three figures in the medal. The dominant figure, standing on the right, is the Algerian, who holds the basin into which the king vomits, as though rendering a kind of demeaning homage. The figure of the Pope, kneeling on the left, is subordinate to Louis in that he is holding the long-handled chamber pot into which the king defecates. Because the text refers to Louis's capitulation to the Pope, however, we may question whether the defecation is merely an act of disrespect. In fact, according to the text (though this is not visible in the engraving) the Pope holds a syringe with which he apparently administered an enema, thus in a sense forcing the king to defecate. The king's feces are therefore a parallel and equally demeaning form of homage.[7]

The imagery in this medal is indeed shocking, and well deserving of the term 'scandalous,' by which it is described in the text. The word 'scandalous' also appears, however, in the description of the exploding bomb: it is said to 'return scandalously to France' before it blows itself up: 'La grosse Bombe que le Roy a fait porter devant Alger, mais qui revient scandaleusement en France & qui se créve' ('The great Bomb which the king had sent to Algiers, but which returns scandalously to France and which kills itself/blows itself up'). The term 'scandaleuse' may mean not only shocking but also shameful, and one interpretation of this text is that the repression of Avignon is the shameful act which destroys the Empire from within, symbolized by the bomb which returns to France before it explodes.

In conclusion, this false medal employs scatology to embody a passionate condemnation of Louis XIV's religious repression. Its message is rendered even more powerful by the literal association between the genre of the medal and the person of the king; as Marin describes, 'medal-money is the body of the king and of the state, present really in their imaginary and symbolic representatives, their portrait and name' (128). The discreet insertion of scatology into a work otherwise glorifying the king was an inspired act of propaganda, guaranteeing that the reader will be caught unawares, and thus affected even more strongly by the negativity of these images. If readers of the present day are still shocked and revolted by this image, we are merely fulfilling the intentions of the Protestant artists whose anger was so profound that it still has the power to shake us after more than three hundred years.

Notes

[1] At the time of its founding in February 1663, this body was called the 'Little Academy,' and was composed of only four members: Amable de Bourzéis, Jacques Cassagnes, Jean Chapelain, and Charles Perrault. It was later renamed the Royal Academy of Inscriptions and Medals, and after the king's death, it became the Academy of Inscriptions and Belles Lettres.

[2] See Menestrier's *Histoire du Roy Louis le Grand*. His previous work included *L'Art des Emblèmes,* and *L'Art des Emblèmes, où s'enseigne la morale par les figures de la fable, de l'histoire et de la nature*. These works, as well as Menestrier's *Traité des tournois* (1669) and *La Philosophie des Images* (1682), have been reprinted by Garland Publishing, Inc. (New York and London), in their series *The Philosophy of Images*, volumes 15–18.

[3] See *Médailles sur les principaux évenements du regne de Louis le Grand* (1702) and expanded edition (1723).

[4] See my articles, '"Ludovicus Heroicus": The Visual and Verbal Iconography of the Medal,' and 'Medals Catalogs of Louis XIV: Art and Propaganda.'

[5] British Library Printed Books, catalog number 1488.dd.6, title page dated 1691.

[6] All English translations are by the author.

[7] The notion of excrement as homage may bring to mind Freud's claim that feces are associated with gold (*Character and Anal Eroticism*, 1908). As evidence, Freud cites a German expression, 'shitter of ducats' (*Dukatenscheisser*), as 'a term vulgarly used for a wealthy spendthrift' (174 n.3). No similar expression appears to exist in Dutch, however, in either the contemporary language or historical dictionaries, according to native speakers consulted by the author. In Charles Perrault's verse fable *Donkey-Skin* (*Peau d'Ane*), published in 1694, there is a donkey which produces gold coins in place of manure. As analyzed by Louis Marin, however, these coins represent the monarch's authority, not his degradation (see *Portrait of the King* 138–65).

Chapter 8

Foolectomies, Fool Enemas, and the Renaissance Anatomy of Folly

Glenn Ehrstine

Stocknarren, Schalksnarren, Geldnarren, Freßnarren: if the Sapir-Whorf hypothesis is true that language reveals the organization and categorization of mental experience in a particular culture, then the experience of fools and folly in Renaissance Germany was indeed broad and varied (Whorf 212 ff.). Whereas the 1568 *Book of Trades* by Jost Amman and Hans Sachs was content to enumerate the four types of *Narren* found above (natural fools, jesters, money fools, and gluttonous fools), Sebastian Brant distinguished over a hundred varieties in the first edition of *The Ship of Fools* from 1494. The success of Brant's work, including subsequent editions in Latin, French, English, and Dutch (Lemmer 175–76), contributed to a detailed taxonomy of foolish types that soon spread throughout Europe. The resulting *Narrenliteratur* flourished in sixteenth-century Germany, aided in no small part by Erasmus's *Praise of Folly* (1509; published 1511).[1] Carnival plays, *fabliaux*, and even sermons employed the fool not only as a representative of human weaknesses, but also as a guileless speaker of truth, capable of exposing the hypocrisies of others.

However, folly was not merely a literary phenomenon. *Narr* could refer to court jesters as well as to 'natural' fools, i.e. the insane (Midelfort 228–76). Similarly, *Narrheit* denoted folly as both mental illness and moral vice. Indeed, the seeming ubiquity of the term or of its Latin equivalent *stultitia* in medical, intellectual, and religious discourse reflects a broad semantic field that defies easy categorization (Kartschoke 142). This has been particularly apparent in responses to one of the better-known studies of historical attitudes towards folly as a mental disorder, Michel Foucault's *Histoire de la folie* (*Madness and Civilization*). While not always denying the validity of Foucault's insights into the development of institutionalized confinement of the deranged, historians such as H.C. Erik Midelfort have noted the author's reduction of *folie* to a predominantly medical term and have striven to produce a more nuanced, empirical view of Renaissance madness that includes folly as a social and theological category (Midelfort 7–10, 229–30; Megill). Regarding *Narrenliteratur* in particular, Ingrid Kasten has noted that Foucault, who asserted that cities actually removed the mentally ill from their communities by placing them on boats, incorrectly took the title metaphor of Brant's work to represent historical reality. Even while acknowledging these criticisms, however, scholars such a Hans-Jürgen Bachorski continue to turn to Foucault for insights on folly and its literary manifestations in Germany.

The following essay examines the melding of moral and medical discourse as found in the physical treatment of folly in four satirical works: the carnival play *Das Narrenschneiden* (*The Foolectomy*), 1536, by the Nuremberg shoemaker-poet Hans Sachs, the dialogue *Eccius Dedolatus* (*The Decornered Eck*), 1520, by Nuremberg Humanist Willibald Pirckheimer, and two rhymed satires by the Franciscan Thomas Murner, the *Narrenbeschwörung* (*The Exorcism of Fools*), 1512, and *Von dem großen Lutherischen Narren* (*The Great Lutheran Fool*), 1520. By taking literary fools as a point of departure for a discussion of Renaissance attitudes on folly, however, I make no claims that the texts are a direct representation of historical medical practice. Indeed, most patients would not have survived the procedures customarily practiced in the period's carnival plays and *fabliaux*, in which physicians took gleeful pleasure in cutting cancerous growths and purging poisonous effluents from the foolishly afflicted. Rather, both operations, foolectomies and what I would like to call 'fool enemas,' reveal contemporary popular conceptions of the physical body as a religious, social, or political construct. For example, Pirckheimer's 'decornering' of the Catholic theologian Johann Eck, Martin Luther's opponent at the Leipzig Disputation of 1519, can be seen as a ritualistic defilement of a religious opponent. Indeed, the numerous acts of purging and cutting found in sixteenth-century *Narrenliteratur* are not unlike acts of purification as described by cultural anthropologists (Douglas), restoring order to both the body of the fool and the society or group threatened by him. Given the carnival context of much of this literature, the stark depiction of physical chastisements can also be taken as a celebration of the grotesque body in Bakhtinian terms, but it is important to note that the physician's treatment of that body reestablishes the socio-political order, rather than subverting it.[2] Foucault himself speaks of 'moral therapeutics of the body' for physical cures deriving from moral precepts (*Madness* 159), and Midelfort notes the opposing, but related tendency of Renaissance physicians to discuss matters of law and theology in medical terms (162).

While these issues are not new, a specific pattern underlies the physical chastisement of fools in the four texts under discussion, revealing a cultural anatomy of folly prevalent in the first half of the sixteenth century. Moreover, this pattern appears unaffected by the cultural changes wrought by the Protestant Reformation. The texts straddle the beginnings of religious reform in their dates of composition, and religious sympathies are evenly distributed among the authors. By 1520, Murner had become one of the leading Catholic polemicists, while Sachs was a staunch supporter of Luther. Pirckheimer, in his way, belonged to both camps: after being threatened with excommunication for his pro-Lutheran authorship of *The Decornered Eck*, the Nuremberg Humanist distanced himself from the new faith and died a Catholic. Among the texts themselves, the *Exorcism of Fools* and the *Foolectomy* do not engage in confessional polemics,[3] while *The Great Lutheran Fool* and *The Decornered Eck* are two of the most effective pieces of Catholic and Protestant propaganda, respectively. In all cases, the curative treatments prescribed by physicians indicate that different types of folly required distinct remedies according to popular conceptions of the body. For superficial follies such as vanity, a simple shearing sufficed to expose lice or other parasites of decorum. Purgation was required for follies involving some form of ingestion, not only of excess food, but also of spurious learning. Lastly, surgical cutting was reserved for deep-rooted follies, most often sin.

In an era that regularly linked mental states to corporeal determinates such as the four humors, it is not surprising that contemporary authors should represent folly as a physical ailment. Melancholy, the most celebrated mental disorder of the Renaissance, was believed to originate in an overabundance of black bile, which together with yellow bile, blood, and phlegm determined one's mood. While melancholy is the privileged subject of most studies of Renaissance temperament (Midelfort 140–81; Klibansky, Panofsky, and Saxl), it should be noted that Galen, the second-century Greek physician who became the leading medical authority of the Renaissance, distinguished *stultitia* and *moria* as types of dementia from *insania* or *melancholia* (Siegel 274). In this paradigm, folly was associated with the sanguine temperament and an excess of blood (185). In all cases, however, the treatment of folly as an illness required some type of physical treatment, as illustrated by contemporary sources. Midelfort relates an anecdote from Johannes Pauli's 1522 *Schimpf und Ernst* (*Jest and Earnestness*), in which a fool, while wrestling with a bystander, receives a wound on the head and regains his senses. The anecdote closes as follows:

> Es ist kein bessere artznei den narren zuo hilff zuo kumen, vnd vnsinnig lüt sinnig zuo machen, dan einem den kopff vff zuothuon, vnd das dämpff daruon gangen. Es ist ein verszlin darüber. *O medici mediam capitis pertundite venam.* (36–7)
>
> There is no better medicine for fools or for restoring the senseless to their senses than to open the head and let the steam escape. There is even a little verse to this effect: … Doctors, perforate the middle vein of the head! (Midelfort 236)

Folk remedies similarly recommended bodily treatments. One cure related by the *Handwörterbuch des deutschen Aberglaubens* suggests restoring a fool to health by wrapping the body parts of a freshly slaughtered young wolf around his head (9: 776).

While the foolectomies and fool enemas employed by Murner, Sachs, and Pirckheimer would not pass as actual medical treatments, they nonetheless draw their effectiveness from the general belief that folly was rooted in some sort of bodily imperfection. Rather than represent the condition as a humoral imbalance, however, these authors consistently treat folly as a type of infestation. Although the ultimate origin of the motif is uncertain, Brant's detailed taxonomy of fools may have lent itself to the anthropomorphization of individual follies, in which they gained volition, intellect, and above all malevolence in their quest to infect unsuspecting victims. In this manner, patients could play host to multiple fools, such as the seven who inhabit the belly of Sachs's protagonist in *The Foolectomy*.

The first representation of distinct follies as contagious pathogens first occurs in the title woodcut to Thomas Murner's *Exorcism of Fools* (Figure 8.1).[4] Here, spritelike *Närrli* fly about with other vermin after being exorcised from a larger fool by the Franciscan monk Murner, who announces in the work's foreword his intention to conjure such fools from all afflicted (1–7). Similar stultifying contagia reside in the Great Lutheran Fool, as seen in the latter work's title woodcut of 1522 (Figure 8.2). Here, Murner, who has now adopted the guise of the cat-like *Murr-Narr* ('Meow-Fool') given him in the Protestant broadside *Karsthans*, extracts the breeches swallowed by the Great Fool, which had been planted in an almond rice dish served at his fictional wedding with Luther's fictional daughter. Upon extraction, other follies come tumbling out of the Fool's mouth together with the breeches. Very little separates these gnat-like follies from tiny demons, which according to popular superstition could cause tumors much like those removed in Sachs's *Foolectomy* (Bächtold-Stäubli 3: 760). Indeed, the title woodcut to Murner's *Exorcism of Fools* portrays an actual exorcism according to local church custom, in which the possessed individual was to be flogged, immersed in cold water, and strangled with a stole (*Narrenbeschwörung* 3). In all cases, treatment of the foolish body required proper diagnosis. This allowed for the adaptation of stock carnivalesque motifs, such as the examination of a urine sample to determine what type of vice a patient had.

Once a diagnosis had been made, the physician could begin with a cure. The least invasive literary treatment for folly occurred as shearing or as a similar cleansing of the external body. Indeed, the treatment of folly in Murner's *Exorcism of Fools* culminates in shearing. After hearing a fool's confession at the end of the book's long parade of follies, Murner assigns the patient the penance of having his or her head shaved:

> Dann wer mir bycht, der nympt die buoß,
> Das er syn har abscheren muoß;
> Thuot er es nit, so weistu wol,
> Das syn har steckt narren vol. (96.49–53)

> For whoever confesses to me, he accepts the penance of shaving his hair. If he doesn't, then you know well that his hair is full of fools.

Equating follies with infectious lice, he notes their preference for finely coiffed heads where they can live undisturbed (although bonnets, we are told, make it hard for them to breathe). The penance of shearing, successfully completed, restores the patient to health on two levels. Physically, it eliminates parasites and, when maintained, prevents any recurrence of vanity based on outer appearance. Yet Murner was an anointed priest and thus able to absolve a confessed sinner. The loss of hair, with the lice of folly, marks the genuine remission of a patient's foolish sins. It would be tempting to equate such a redeeming hair loss with a monk's tonsure, except that Murner informs us that many monks devote themselves to fashion just as religiously, or non-religiously, as the laity (96.41–6).

Figure 8.1 Thomas Murner, title woodcut from *Narrenbeschwörung* (1512). Courtesy of the Staatsbibliothek zu Berlin Preußischer Kulturbesitz, 19 ZZ 4100.

Figure 8.2 Thomas Murner, title woodcut, *Von dem grossen Lutherischen Narren* (1522). Courtesy of the Bayerische Staatsbibliothek, Rar. 870.

Given the ubiquity of scatological references in the satirical literature of the period, the most common treatment of physical folly seems to have been that of a humid brain, denoting dullness, stupidity, and somnolence (Midelfort 159). In at least one instance, contemporary physicians seem to have shared Murner's concern with hair as a sign of mental health, interpreting straight hair as the mark the fool enema. A woodcut from Murner's *Great Lutheran Fool* gives graphic depiction to such a procedure (Figure 8.3). Murner has brewed a special laxative from the leather of the *Bundschuh*, symbol of the insurgent peasants' movement in south-west Germany during the late fifteenth and early sixteenth centuries. This distilled essence of rebellion produces a small fool from the Great Fool's bowels. The woodcut marks him by his *Karst* ('mattock/hoe'), and the text reveals that this denizen of the duodenum is Karsthans, the quintessential Lutheran peasant in early Reformation pamphlets. As the title figure of the eponymously named broadsheet of 1521, Karsthans's lack of artifice allows him to see through the supposed sophistry of Catholic theology, represented by Murner as the *Murr-Narr*. By linking Karsthans to feces in *The Great Lutheran Fool*, Murner avenges himself for this dehumanization. Furthermore, the association of the purged fool with peasant rebellion suggests that Karsthans is not as God-fearing and upstanding as he might appear. Indeed, when Murner asks the Great Fool what Karsthans has been doing down there, he replies that Karsthans has been hiding from the imperial edict against authors of polemical tracts:

> Als bald er höret dise mer,
> Wie das der keiser zornig wer,
> Wolt er da ylends nim beharren
> Bei den andern grosen narren
> Vnd hat sich in mein ars verborgen. (2694–98)

> As soon as he heard the news that the emperor was enraged, he suddenly no longer wanted to stay among the other big fools and crawled into my arse.

Karsthans's willingness to take up residence in such cramped quarters thus seriously discredits his authority as spokesperson for the new faith.

In the specific context of Reformation polemic, Karsthans's follies are not universally applicable to all of humanity, i.e. they do not embody a general vice or sin found in human nature. Nonetheless, his contamination of his host parallels other cases of foolish infestation ending in enemas. As the title figure of the anti-papal pamphlet *Karsthans*, the peasant symbolizes the Protestant propaganda ingested by the Great Fool. To use a German idiom, 'der Narr hat's geschluckt,' that is, the Great Fool has gullibly swallowed Lutheran theology hook, line, and sinker, and it now festers in his digestive tract. In this regard, any form of learning that has been taken in intellectually is, under the proper circumstances, a candidate for purgation. Thus the pamphlet series the 'Fünfzehn Bundesgenossen' ('The Fifteen Confederates') by Johann Eberlin von Günzburg, which the Great Lutheran Fool has also consumed, similarly finds its way into his belly, where the confederates become *buchgnossen* ('belly confederates') (804–5), Murner's pun on the title's original *buntgenossen*. Protestants could equally employ fool enemas against Catholics, as we shall soon see in *The Decornered Eck*.

Yet even purging is insufficient for learning that has become an end in and of itself, to no greater betterment of others. Self-serving erudition joins the ranks of the most sinful follies, which can be removed only by the knife, as occurs in both Murner's *Exorcism of Fools* and Sachs's *Foolectomy*. As its fifth chapter, Murner's work of 1512 contains a section devoted to 'Skinning Learned Fools' ('Gelerte narren schinden'), based upon a woodcut first created for Brant's *Ship of Fools* (Figure 8.4). The illustration originally depicted an episode from Ovid's *Metamorphoses*, namely the punishment of Marsias, who had challenged Apollo to a musical competition and lost. Murner, however, interprets the image literally. To a learned fool who has questioned his authority and challenged him to a disputation, Murner replies that his opponent is a perfect candidate for a foolectomy:

> Myn fryheit sag ich in vor an,
> Die ich von vnserm keiser han
> Erholet, maximilian,
> Der mirs zuo wurms vff einem tag
> Erloubt, das ich üch schinden mag,
> Durch allen körper hin vnd wider
> Suochen, mercken alle glider,
> Ob yendert ich ein örtlin findt,
> Do narren inn verborgen sindt.
> Jr sindt mit worten also kluog,
> Das kein beschwörung ist genuog;
> Mit worten halt ir alzyt fuoß,
> Darumb ich üch hier schinden muoß. (5.81–93)

> I first announce to learned fools that I have obtained license from Emperor Maximilian, granted during a Diet at Worms, that I might skin you, searching back and forth throughout the whole body, taking note of all limbs, to see whether I might find a spot where fools lie in hiding. You are so clever with words that verbal exorcism alone is insufficient. With words, you always manage to keep your footing; that's why I have to skin you.

According to Murner, surgical cleansing is the only recourse with learned fools, since they are so rhetorically skilled that verbal exorcism has no effect. And since the educated belonged mainly to the clergy, Murner argues in a pre-Reformation critique of ecclesiastical corruption that cutting alone can eliminate the sophistry with which many clerics justify their gluttony, unchastity, and accumulation of benefices.

Hans Sachs's play *Das Narrenschneiden* of 1536, revived by Johann Wolfgang von Goethe for two productions in eighteenth-century Weimar, is perhaps the best-known example of the surgical treatment of folly. Like Murner as *Narrenbeschwörer*, the surgeon in Sachs's *Foolectomy* must also treat gluttony and unchastity in curing his patient, as well as the five remaining deadly sins. These seven follies are located in the patient's belly, and their subsequent delivery, as

Figure 8.3 Thomas Murner, Karsthans' 'enema,' *Von dem grossen Lutherischen Narren* (1522). Courtesy of the Bayerische Staatsbibliothek, Rar. 870.

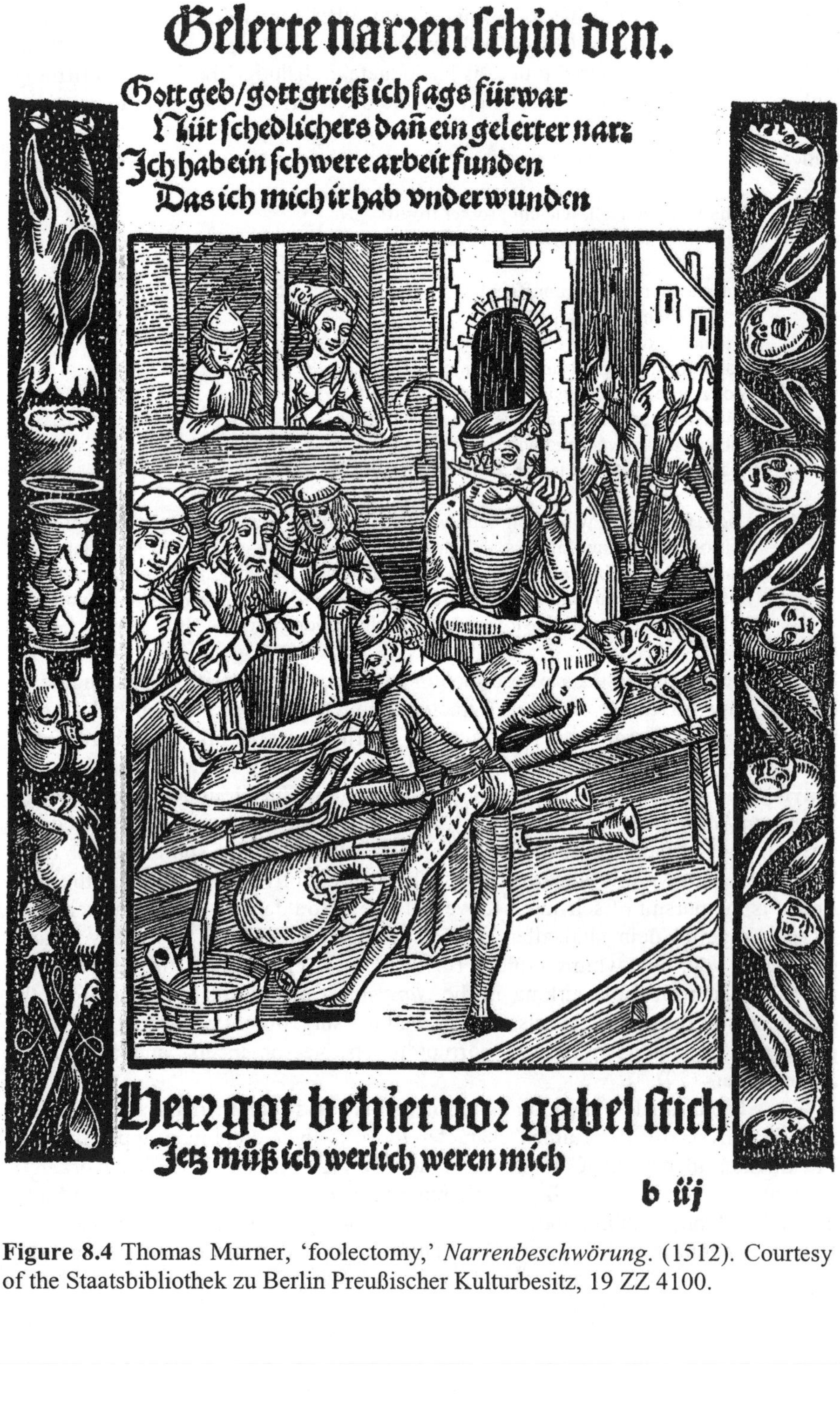

Figure 8.4 Thomas Murner, 'foolectomy,' *Narrenbeschwörung*. (1512). Courtesy of the Staatsbibliothek zu Berlin Preußischer Kulturbesitz, 19 ZZ 4100.

noted by Ralf Erik Remshardt, resonates with other carnivalesque motifs such as that of the pregnant man (345–49). However, the surgeon's prescription of beer, buttermilk, and two hundred prunes is inadequate to dislodge the fools, confirming that only surgical removal is sufficient for the most serious vices. As the physician cuts away the tumors one by one, he comments on their appearance, such as that of pride:

> Schaw, mein gsell, wie ein grosser tropff!
> Wie hat er so ein gschwollen kopff!
> ...
> Der narr hat dich hart auffgepleht.
> Er übet dich in hoffart stät.
> Wie hat er dich so groß auffplasen,
> Hochmütig gemacht ubermassen,
> Stoltz, üppich, aygensinnig und prechtig,
> Rümisch, gewdisch, samb seist du mechtig!
> Nicht wunder wer, und wilt dues wissen,
> Er het dir lengst den pauch zurissen. (9.4–5, 10–17)

> Look, my friend, what a big fellow. What a swollen head he has!... This fool made you overblown and always accustomed to pride. How greatly did he inflate you, making you hubristic beyond measure, arrogant, haughty, pigheaded, pompous, boastful, and extravagant, as if you were mighty! It's surprising, should you care to know, that he didn't tear your belly to shreds long ago.

As the patient's other sins are described in anthropomorphic terms as having heads or being haggard, they were likely depicted during the performance of Sachs's play by small puppets similar to the *Närrli* in Murner's woodcuts (Remshardt 336). Indeed, a puppet-like fool emerges from the patient's belly in the woodcut accompanying the original edition of Sachs's play, reproduced in Richard Schade's *Studies in Early German Comedy* (95). After the seven sins have been excised, the surgeon must finally remove the *Narrennest*, which is the source of all other follies. As with Murner's learned fools, the root of folly, as revealed by the physician, is the pursuit of self-interest over all else: 'Von dem kamen die narren dein, / Das dir gefiel dein sinn allein' ('Your follies came about because you took pleasure in your own whims alone') (16.12–13). As Schade notes, the discussion of the as yet unborn fools lurking in the *Narrennest* also shifts attention from the deadly sins to the social aspects of folly: had the surgeon not removed them, these proto-fools would have become corrupt jurists, necromancers, and alchemists, among others (79).

The ultimate indication that foolectomies and fool enemas each addressed specific, physically determined character flaws is their combined application by Pirckheimer in his anti-Catholic dialogue *The Decornered Eck*. The work is a veritable *tour de force* of all possible physical chastisements of the foolish body, enacted with one goal in mind: to discredit as thoroughly as possible Johannes Eck, Luther's opponent at the Leipzig Disputation of 1519 and subsequently among his most vocal critics. At the beginning of the dialogue, Eck is suffering excruciating pain from an unidentified illness. At the advice of his sycophantic friends, he sends a witch to the conservative theologians of Leipzig, who had boasted of Eck's supposed victory over Luther, to obtain a physician who might help him. After examining his patient's urine, the surgeon announces that Eck is suffering from *stultitia* and will surely die unless he undergoes a drastic treatment for his ills. Eck has no choice but to suffer a series of violent remedies, which parallel the cures assigned by Murner or Sachs. Following a beating, Eck, too, is shorn of his hair, revealing the 'lice' of syllogisms and sophistry. Afterwards, one half of his forked tongue is removed before a laxative causes him to evacuate his digestive tract at both ends. He must first vomit up his theological commentaries along with his doctorate in canon law; afterwards, he voids his bowels of indulgences along with the money that he has allegedly taken from the Fuggers of Augsburg for his theological defense of usury. Lastly, a thorough foolectomy begins, in which Eck's skin is removed to reveal the tumors of boastfulness, self-love, wastefulness, hypocrisy, flattery, deceit, and envy, among others. Before wielding the knife, the surgeon compares Eck's predicament to that of Marsias, whose image had served as the original source for the *Exorcism of Fools* woodcut on skinning learned fools. The unkindest cut comes last of all, when the surgeon castrates his patient. In those procedures where a comparison is possible, only the act of shearing seems to differ slightly from the earlier examples of foolish discipline: the penance of the bald head in Murner's *Exorcism* addressed vanity and the outer appearance of the body, whereas the shearing of Eck exposes his intellectual deficiencies. As with the Great Lutheran Fool, however, the purgation of Eck's belly and bowels rids the patient of metaphorically ingested matter, such as Eck's *Theologia negativa* or the 'indigestible' coins of the Fuggers. Cutting is reserved for character flaws, such as *philautia*, Eck's self-love, which has grown under the skin.[5] That Pirckheimer successfully linked Eck to the foolectomy motif is demonstrated by Eck's subsequent appearance in the pamphlet *Ain schöner Dialogus. Cuonz und der Fritz* (*A Pleasant Dialogue between Kunz and Fritz*), 1521, likely written by Urbanus Rhegius (Bentzinger 18–19). Here, the peasant Fritz states that Eck's wound has reopened and that his foolish tumor is larger than it ever was, threatening to either kill him or infect others:

> Als er am narren beschnitten ward, der dann so groß in im gewachsen ist und überhand genomen, daß sich ain groß wunder diss dings in im erhebt hat, zoch er gen Rom. da hat er sich nit wol gehalten. in der selben erznei ist die wund wider auf brochen und ist der nar vil größer worden dann er vor was. und wirt er in nit bald wider schneiden laßen, so het ich sorg, er müest daran sterben oder er wurd Bollingen, da er iez ist, das ganz closter mit narren besetzen. (Schade 2.124)

> Now, when he was operated on his Fool, which had grown so large in him that it had gotten out of hand, so that it was quite a miraculous thing which he issued forth, he went to Rome. There, he wasn't well for long. Treated with the same medicine the wound opened up again and the Fool grew much larger than before. And if he doesn't have it cut soon, I'm worried that he will die of it or that he

> could infect the entire monastery of Bollingen, where he is at present, with Fools. (Remshardt 339)

Although Pirckheimer leaves Eck a cured man at the close of his work, the dialogue suggests that Roman theology is itself a disease, recontaminating Eck and infecting all who come in contact with it.

Throughout these examples, medical discourse merges with moral imperatives in cleansing both body and soul. The physician's initial examination of a urine sample produces pronouncements on the patient's lack of moderation in any number of activities. An infestation of lice on the head marks the patient as a fool of fashion and leads to the condemnation of vanity. The maintenance of the digestive tract requires the avoidance of impure nourishment, including that of false learning or theology. Finally, the presence of tumors denotes the habitual repetition of any unhealthy behavior, resulting in deeply entrenched vices and even sin. In this last instance, the patient can no longer affect a cure on his own through a change in diet or hygiene. The restoration of health is impossible without the intervention of the physician, and it is no accident that the *medicus* in all four works takes on the role of the confessor, forcing the fool to enumerate his depravities before surgical treatment can begin. An investigation of further texts is, of course, necessary to determine the extent of such foolectomies and fool enemas. Given the popularity of the physician as a stock character in early carnival plays, an examination of the corpus of fifteenth-century German *Fastnachtspiele* would be prerequisite in exploring the origins of the motif prior to Murner. Still, the relatively homogenous representation of the foolish body by Murner, Pirckheimer, and Sachs indicates that contemporary authors constructed that body in accordance with broad intellectual, moral, and physical values. The violation of those values, these authors implied, endangered not only the health of the individual, but also of German society as a whole.

Notes

[1] On *Narrenliteratur* in general, see Könneker, Bachorski/Röcke, Groß, and Kartschoke. Translations throughout are mine unless otherwise indicated.

[2] For more recent critiques of the populism inherent in Bakhtin's theories, see Stallybrass and White, and Flanigan.

[3] There is no evidence for Ralf Erik Remshardt's suggestion that the surgeon in Sachs's play represents Martin Luther, much less that the play responds to the Anabaptist coup d'état in Münster. Cf. Remshardt 338, 352.

[4] The appearance of anthropomorphized fools as early as 1512 belies Remshardt's suggestion that the *Narren* of Sachs's *Foolectomy* have their roots in Murner's *Great Lutheran Fool* or the *homunculi* of Paracelsus. Cf. Remshardt 339, 342.

[5] Könneker notes the relationship of Eck's cutting to the foolectomy motif, but also points to classical sources such as Lucian, which may have influenced Pirckheimer as well (Könneker 162).

Chapter 9

Holy and Unholy Shit: The Pragmatic Context of Scatological Curses in Early German Reformation Satire

Josef Schmidt, with Mary Simon

Profecto, mich verwundert daß Gott so hohe Artzeney in den Dreck gesteckt hat.[1]

The German language has an unusually high elective affinity for scatological language. Folklorist Alan Dundes put succinctly what has been common knowledge all along:

> In German folklore, one finds an inordinate number of texts concerned with anality. Scheisse (shit), Dreck (dirt), Mist (manure), Arsch (ass), and similar locutions are commonplace. Folksongs, folktales, proverbs, riddles folk speech – all attest to the Germans' longstanding special interest in this area of human activity. (9)

For example, the standard modern colloquialism for cheating is still *bescheissen* ('to shit upon'). The expression has a strong pedigree; its connotative etymology can be found in many works in the vernacular of the Gutenberg era. The first popular bestseller, Sebastian Brant's *Das Narrenschiff* (*Ship of Fools*), 1494, delineates in section 102 under the heading 'Von Falsch und Beschiss' ('Of Falsehood and Cheating') two particular areas of the metaphorical expression that are significant in our context. The first is the obvious connection to cheating in business, be it with false coins or dirty dealings; the other is a condemnation of 'alchemy' used in the narrow sense of brewing late medieval patent medicines with questionable ingredients and processes. This is no casual coincidence: money and medicine had a strong connection to human excrement in late medieval times. And in turn, scatological language permeated the popular discourse of that period because it was far more manifest and visible in public and private life at that time than it has been in more recent periods of modern history.[2] Before turning to the scatological evidence in early German Reformation satire I shall demonstrate briefly what has to be included in an historical understanding of expressions like *bescheissen*.

One of the most widely known pre-Reformation anti-clerical texts is the German picaresque collection of *Till Eulenspiegel* stories (after 1500). This trickster who led burghers, farmers, and priests by the nose reserved his scatological triumph for the deathbed.[3] Anecdote 92, 'Wie Ulenspiegel sein Testament macht, darin der Pfarrer sein Händ bescheiß' ('How Eulenspiegel makes his last will and testament and a priest gets shit on his hands'), narrates how he lets a priest know he wants to make peace with the church. He indicates that he will leave his money to the priest provided the latter will give him the sacrament of the dying. But when the greedy priest gropes for the hidden gold coins, he finds them in a dark closet, and only too late does he realize in his hasty quest that they are at the bottom of a full chamber pot. Sigmund Freud, therefore, was not the first to make the connection![4] Here we have the anti-clerical triad of money, shit, and religion that was to fully explode into the satirical Reformation discourse as holy and unholy shit.

The association between money and feces had already developed into a whole array of images by the end of the thirteenth century. The root of this phenomenon is generally assumed to be the expression of anxiety and fear in the face of a profit economy where coins replaced a commerce of bartering. The distrust towards this new practice expressed itself in many ways. From the Devil luring the greedy with gold coins which then would turn into human waste, to apes shitting gold coins, there was a definite anal tradition of imagery in gothic manuscripts. And the 'association of filth and fecal matters with one's enemies' took special significance when it became a common theme for associating the foul-smelling creature of the Devil with moneylending Jews (Little 53 *et passim*).

To understand the actual curses in early Reformation times it is appropriate to contextualize the most dominant traditions of scatological expletives in the late medieval period and during the Renaissance. These expletives surfaced as part of the religious controversy in the sixteenth century.

1. Shrovetide plays often refer to scatological matters; they were a natural genre of folk humor for articulating what must have been an important feature of popular colloquial speech of that time. These plays were staged by burghers before Lent. The plots were based on everyday comical situations. Standard stereotypes included the shrew, the cuckold and vices associated with particular professions, e.g., the garrulous tailor (Spiewok 11). Early Reformation satire borrowed liberally from this form of ridiculing one's adversaries as will be shown later in this essay.
2. Another source that became very popular during the Reformation were collections of *facetiae* (*Gesprächbuechlin*) that found their way from the oral tradition into print. Their primary aim was entertainment; and since the prefaces often mention that these stories were told in inns and public bath houses, one is not surprised that bodily functions form an essential part of this often very earthy humor.[5]

3. A verbal source of very differentiated popular scatological language that is often neglected can be found in popular medicine. The *Dreckapotheke* – medicines and cures using excrements and other body secretions – was widely used in German-speaking countries by the common people. The Luther quotation at the beginning of this essay attests to that feature.
4. Another powerful aspect can be traced back to medieval piety: the excremental imagery associated with evil and the Devil.[6]

I will discuss these four aspects in reverse order to show the historical confluence of fecal traditions into a torrent of scathing scatological vituperation during the early Reformation. The representative examples are two particular scenes of one of the most prominent Reformation satires, Thomas Murner's *Von dem großen Lutherischen Narren* (*The Great Lutheran Fool*). It is the perfect German illustration of Mikhail Bakhtin's carnival culture thesis that the ambivalent laughter culture, with the arrival of the Renaissance, gave way to a new kind of grotesque mockery. One notes that medieval laughter, suspending the rigid rules of social conventions, became derisive, mocking and even abusive. At the same time, the ironic judgment of the satirist moved away from earthy folk humor.[7] Murner's satire reflects these two stages in the early period of the Reformation. The rather extensive verse epic oscillates between pure carnival festive laughter and targeted vicious attacks such as the *ad hominem* against religious opponents. It truly represents a watershed in satirical Reformation discourse.

(4) According to David Williams, a specialist in medieval monstrosities, 'The human body is the primeval matrix of all the chief figures and analogies by which human language seeks to understand reality...' (176). No wonder, then, that in medieval piety, the Devil and hell were associated with scatological discourse. A *locus classicus* of the innermost hell is situated below the Devil's tail. A canonical text for the accompanying imagery of this strain of medieval piety is Chapter III, 21 of the revelations of Mechthild of Magdeburg, *Das fliessende Licht der Gottheit* (*The Flowing Light of the Godhead*). Luther's knowledge of the Devil and his works is directly based on this text as he refers to the nun in his *Tischreden* (*Table Talk*) (258,b). Two central images found their way into Murner's satire: 1) that mortal sinners are devoured and spat out by the Devil, and 2) particularly devious sinners have to go all the way through his digestive system until they exit through his anus. The context of Luther's preoccupation with the Devil is his recipe on how to deal with temptation. Erik Erikson's discussion of Luther's anal fixation, interesting as it is, commits a Classical Freudian error of individualizing concepts that are, in reality, individual adaptations of very ancient traditions.[8] Luther recommends the following remedies for hellish temptation. In order to ridicule the Devil, one should address him thus: 'Devil, I also shat into my pants, did you smell it, and did you record it with my others sins?' (*Tischreden* 267,b). A more concise recipe of Luther's is that Classical comment that most German speakers associate with Goethe's famous exclamation in *Goetz von Berlichingen* (3:4): 'Leck mich im Arsch' ('Lick my ass') (*Tischreden* 269,a) or in a more elevated version: 'Kuesse mich auffs Geseß' ('Kiss my behind') (278,a). Luther

also provides a female version of how to deal with debasing the Devil: a woman responds to temptation by baring her behind, farting profusely and telling him that this should be his pilgrim's staff for his journey to Rome in order to get a proper indulgence from the Antichrist (267,b). The well-thumbed pages of the original print the author consulted (Zentralbibliothek Zürich WZ 23) attest to the popularity of this kind of information.

Luther displayed a very direct and uninhibited attitude toward the use of urine and feces. From the depths of hell to heavenly heights, this theological shit-spreader provided the scatological with a fervor never heard or read before. His often very coarse and crude imagery has delighted generations of Lutherans right up to our time. His tirades against the pope especially indulge in this type of fecal imagery, as the following examples from a recent anthology can demonstrate. When pitying a corrupt Rome, he laments the city that was re-created with the blood of martyrs, 'until the Devil shat the pope, his very own shit, on it.' And likewise, 'The pope is the cuckoo, who devours the church's eggs and then craps plenty of cardinals' (Krumbholz 13f.). But the Lord of Flies is also mentioned in other Lutheran contexts such as 'Here we have the cattle in its stable, said the Devil, and drilled a fly right into his mother's ass' (Krumbholz 32f.).

(3) There was, however, a strong positive connotation to animal and human excrement with roots going back to antiquity in the form of the German *Dreckapotheke*. This term is commonly translated as 'filth pharmacy' or 'dirt pharmacy' which is an inept and even misleading rendering. The appropriate term is 'fecal pharmacy' both for linguistic and substantive reasons. In early High German, most words had a much broader range of meanings than in later periods. The term *Dreck* is an excellent example of this phenomenon. According to Grimm's *Wörterbuch* (2:1352), the most comprehensive German dictionary, the term designated the following three basic literal meanings: 1. *unreinigkeit/schmutz* = unclean matter/dirt; 2. *weiche erde...thon, lehm* = soft soil/clay; 3. *mist, excrementum* = manure, excrement. But figurative connotations of the term might be as profound as referring to the material from which man was created. The English speaker can easily get a sense of these connotations when s/he compares the nouns and the adjectives of 'soil/soiled' or 'dirt/dirtied.' Especially in the first case there is a huge gap between the technical label for earth and the shifting adjective meaning for 'unclean/dirtied,' e.g., for a diaper.

On substantive grounds, there is the fact that the *Dreckapotheke* included not only excrement, but also secretions of the human body such as menstrual blood or sweat, and animal matter. The Classical text, the *Dreck-Apotheke* by Kristian Frantz Paulinus, appeared first in 1696, but its content dates back to much earlier times – e.g. 'urine inspection' as a diagnostic tool was already firmly established in the middle ages (Wallis 272f.). It is a highly comprehensive and detailed manual. An example is found in the treatment for cancer of the breast(s), Chapter XI: four pounds of cow dung, six distinct herbs and the remains of a dozen crabs should be distilled into a liquid that should then be regularly applied to the breasts until healing occurs. A multitude of variations – the dung of goats, geese, etc. – supplement the original prescription with honey, wine, etc. Numerous references to individual doctors demonstrate that the *Dreckapotheke* was indeed a widely practiced form of medicine in popular culture. A popularized version of another canonical text, the *Thesaurus pauperum* – written in the fourteenth century but also dating back to much earlier times – prescribes a much simpler cure for tumors and female itches (*Contra tumorem et pruritum virgae*, 65:195). In addition to cold water, *lavetur / Urina propria purgando cum benedicta / Urina*... (The appropriate part of the body was then treated with the patient's own urine in addition to blessed urine...). The important feature to remember is that excrement occupied a very vital dialectic function both in cures and in curses. A modern source lists fecal and urinary medication still used in the nineteenth century as remedies in tabulated form, curing just about everything from leprosy to syphilis (Monestier 198). The basic mechanism was the ancient form of fighting an existing physical or psychological curse by applying the repulsive matter to the object affected by it. One of the proponents of the *Dreckapotheke*, Becher, proposed in his eulogy on the *stercus humanum*: 'Die große Schmertzen so durch Hexerey gemacht / Die werden durch den Koth des Menschen weggebracht' ('Great pains induced by witchcraft / Can be removed through human excrement,' quoted in Bächtold-Stäubli: col. 342). For the postmodern student of scatology there remains the sad fact that contemporary users of the four-letter word *shit* (or *Scheisse* = 2x4!) employ an incredible paucity of meaning when articulating themselves on an essential fact of human life.

(2) Many of the *facetiae* (immortalized in high literature by Boccaccio in the fourteenth century and collected in the sixteenth century) had a place as illustrating examples in popular homilies. No wonder, then, that under the guise of entertainment there often lurks a strong note of moral condemnation. But the *facetiae* also offer a rich source of information about everyday life in late medieval times and the Renaissance. A typical example is the 153rd story of Johann Pauli's *Schimpf und Ernst* (*Jest and Earnestness*), 1522. The motif is of the traveling scholar who cheats illiterate people ('ein lüthbeschisser') with his learning. He proposes a cure for a poor woman with bad eyesight by writing a blessing on a piece of paper that she should wear around her neck. But when she asks the priest to read the actual text, he laughs. The cleric adds insult to injury by spelling out the curse that the young scholar had written 'Der hencker stech dir die augen uß, und der tüffel scheiß dir in die lücken' ('May the hangman cut out your eyes, and may the Devil then shit into the holes,' quoted in Bobertag 104). The actual curse is here diluted as an insult of the 'scholar' by mocking his social inferior.

The handbook for conducting witch trials, Heinrich Kramer and James Sprenger's odious *Malleus Maleficarum,* contains a host of sinister anecdotes about witchcraft that deal mainly with sexual imagery. Several of these mention incidents of heretical activity with names, places, and defining circumstances. There is, however, the odd scatological reference which deals, not surprisingly, with blasphemy. In part II, question 1, Chapter 1, there is reference to a witch, 'protected by secular law' (96), who translates the priestly 'Dominus vobiscum' ('The Lord be with you') during mass 'in the vulgar tongue' into 'Kehr mir die Zung im Arss umb' ('Screw your tongue in my arse'). This represents the other extreme of anecdotal report: a scatological utterance that would land the unfortunate perpetrator in the hands of the Inquisition. The genre of the *facetiae* illustrates very clearly the range of scatological language and imagery of that period extending from the hilarious to the demonic.

(1) Shrovetide or carnival plays were, by their very nature, rooted in the here and now. The predominant celebration of earthly life centered strongly on sexuality and the fecal idiom (Catholy 41). The central figure was most often the peasant, or rather a caricature of a peasant as seen by the burghers and city dwellers. And it is no wonder that the ideal plot was that of a wedding where sexual and fecal jocularity could reign freely. When discussing one of the most prominent early Reformation satires, all these functions should be kept in mind. Thomas Murner's *Conjuring up the Great Lutheran Fool* is an eclectic satire rather carelessly thrown together but it receives its vitriolic power from late medieval forms of exposing human vices through the central image of the reprehensible fool.[9]

The verse epic falls into two heterogenous parts, mixing metaphors of conjuring up and summoning the reformer, giving birth, and defecating him, thereby launching into a vicious attack on Luther. The combination of digestive, anal, and genital phrases oscillates between Shrovetide soiled garment humor (Catholy 44) and grotesque condemning curses.[10] The central image is a clever combination of exorcism and medical purgation. The first part introduces a gallery of early Reformation proponents, some of whom had already engaged Murner in public controversy. They enter the literary stage in Shrovetide-like fashion, engaging in farcical dialogue. Murner dispenses with them one by one and covers them with acid scorn. They are conjured up from all the orifices of Luther the fool – but they also creep out of his pockets, among other things. The second part depicts a scene where Luther tries to marry off a daughter to Murner in order to mend fences; however, since she turns out to be covered with scabies and also has other undesirable traits, Murner forgoes the proposal.

The verses are peppered with scatological expressions in a most explicit manner. One of the favorite images being the peasants' 'verschissne bruoch' ('the thoroughly soiled swaddling clothes'). The woodcut accompanying Murner's condemnation of one of the most popular early Reformation fictional figures, Eberlin v. Günzburg's liberation peasant Karsthans, is perhaps the most telling example (2631ff.; Schmidt 80ff.) of the central scatological image (Figure 9.1). In the woodcut, the Lutheran fool is given a strong purgatory potion and eventually defecates on the culprit. The Franciscan friar holds a vessel loaded with urinary meaning. It is a 'jordan,' a glass flask used by alchemists and pharmacists to analyze urine. The shoe emblem is the well-known symbol of the German Peasants' Confederation that would later be employed in the devastating peasants' war. The analogy drawn by a bestialized Murner (he liked to portray himself as a tomcat because his enemies used this caricature based on a pun on the Franciscan's name: 'Murrnarr' = 'Murnerian fool') is direct. That Reformation pamphlets should be unmasked and situated for what Murner thinks about them is considered nothing less than a shameful waste.

Figure 9.1 Thomas Murner, Karsthans' 'enema,' *Von dem grossen Lutherischen Narren* (1522). Courtesy of the Bayerische Staatsbibliothek, Rar. 870.

At the wedding party, Murner serves a kind of peppered stew that Luther finds indigestible. He goes so far as to guess that it must have been prepared from the Devil's skin (4144). Murner publicly shames the wedding party by confessing that he simply took one of the insults of his adversaries, wore soiled undergarments, and turned it into a dish. He scorns them for not having been able to identify the object in question before devouring it. The scatological metaphor for theological ignorance is coarse but clear.

When facing the stark and strong excremental imagery of early German Reformation invective, it is important to keep its popular roots in mind. While some of it already was targeted as a satire and grievance about ecclesiastical abuse in earlier times,[11] most of it exploded during the first decade of the German Reformation in full force. Shrovetide or carnival episodes were transformed for the new cause. Critical anecdotes became condemning narrations. Accordingly, popular medicine provided powerful images for a fight with no resolution. Demonic and satanic images of popular piety were harnessed by Catholics and Lutherans to illustrate their respective causes.

The scatological bent of early Reformation satire is strong and pungent. It was lifted into print from a variety of sources: medieval images of negative (diabolical) piety, patent cures, colloquial invectives, and scenic entertainment. As the controversy raged on, these images became entrenched in the new religious discourse reflecting the direct relationship people at that time had to the world of scatology and its sources. But, as Bakhtin described in another context (*Rabelais and his World),* the ambivalence of carnival and laughter now gave way to unequivocal irony, sarcasm, and derision. The many instances of actual abusive degradation in images of the lower stratum of the body did not provide the satirical images of grotesque realism with destruction and regeneration, but ended rather in coarse and vituperative condemnation, and, as mentioned at the beginning of this essay, Reformation satire and invective undoubtedly provided a powerful contribution to the scatological layer of the German language.

Notes

[1] 'Indeed, I am pretty perplexed why the Lord put such valuable medications into human excrement' (Martin Luther quoted in Bächtold-Stäubli: col. 342).

[2] This obvious observation has not fully entered the historical understanding of that period. Cf. Fernand Braudel's *Civilization and Capitalism, 15th-18th Century* (1979), where the noted historian of mentalities barely touches on issues of sanitation, etc. On the other hand, cf. Dominique Laporte's *History of Shit* (1978), particularly Chapter 1, quoting ordinances dealing with the removal of human excrement in sixteenth-century France.

[3] Throughout the collection, stories abound where the trickster carries out his revenge against hostile burghers by defecating and using his excrement, e.g., in

No. 77 he spoils a party in an inn to which he is not invited by fanning the stench of his feces onto the merry guests.

[4] V. 'Introductory Lectures on Psychoanalysis' (XX. The Sexual Life of Human Beings) 315. A more extensive treatment of the equation infant's feces = gift = money, turning into adult avarice, is presented in his article 'Über Triebumsetzungen, insbesondere der Analerotik' (On Transformations of Instinct as Exemplified in Anal Eroticism) (1916–17).

[5] On *facetiae* collections, see Barbara Bowen's article in this volume.

[6] Cf. Inglis 81: 'The clerical culture drew upon Christian symbolism of an ancient pedigree to illustrate the theological premise of the debased nature of humankind, such that the wretched human race was likened to a heap of excreta, and the human body itself was nothing but a corrupt husk which harbored such foul materials.'

[7] While using Bakhtin's broad categories, I find his subcategories insufficient; cf. *The Bakhtin Reader* 196ff. They read like a typical formalist catalog, being inconsistent and incomplete. In the present case: he does not include medieval traditions of piety or of popular medicine in the Rabelais analysis.

[8] Cf. Erikson, Dundes 59–61 and Schmidt 35 (containing several pertinent illustrations).

[9] Murner had written quite a few successful traditional satires on human vices before venturing into this polemical Reformation invective. Because he had also criticized certain church abuses, he was first considered a potential convert to the Lutheran cause. But with this satire he publicly showed his colors as a proponent of the 'old church.' Apart from his skills in publishing in the vernacular – he was one of the few Catholic polemicists, who could hold his own with Luther – he was also a skilled and trained theologian. But because of his garrulous character, immersing himself in ever more personal controversies, he gradually faded into the background during the unfolding Reformation events.

[10] For a comprehensive treatment of this phenomenon, see my article on 'Luther the Satirist.' I would also like to express thanks to my colleague at McGill, historian Faith Wallis, who suggested a few pieces of information and prevented me from reporting some errors.

[11] Like the Eulenspiegel-episode mentioned at the beginning of this article.

Chapter 10

Expelling from Top and Bottom: The Changing Role of Scatology in Images of Peasant Festivals from Albrecht Dürer to Pieter Bruegel

Alison G. Stewart

During the first half of the sixteenth century, the earliest visual representations of peasant festivals in European art were produced in Germany.[1] These works, all prints, showcase peasants expelling their drink with the result that art historians today, nearly 400 years later, have described these prints as gross and indecent. In their revulsion and distancing from sixteenth-century Germany's insistently colorful visual and verbal vocabulary, art historians of Northern European art appear to have stressed both the values and preferences of their own twentieth-century culture and that of the sixteenth-century Netherlands rather than those of the society that produced them – sixteenth-century Germany. This preference for clean peasant festival images, ones that are scatologically free, underscores the art historical preference for Netherlandish art that includes Pieter Bruegel's paintings and prints of peasant festivals over the German prints, which originated the theme.

Although Germany produced the first images of peasant festivals in European art, it is the later Netherlandish prints and paintings that have been studied and showcased in art historical literature, despite recent studies of the German imagery by Margaret Carroll, Keith Moxey, and myself. The makers of prints turned their skills to images centered around peasants feasting and celebrating first in the form of prints in Germany, at the home of its most celebrated artist, Albrecht Dürer, at Nuremberg, and then a half century later as paintings produced to the north and west in Flanders and Antwerp, with Pieter Bruegel.

In this essay I study the earliest peasant festival imagery that was designed by Sebald Beham at Nuremberg during the 1520s and 1530s. I investigate the imagery produced as large-scale woodcuts, drinking, and related issues within the cultural contexts of Lutheran Nuremberg during the 1520s and 1530s soon after the adoption of Lutheranism in Nuremberg in 1525. I then trace the production of peasant festival images from Nuremberg to Frankfurt am Main, where Beham continued designing peasant festival prints which he reinvented as small engravings in series published in the late 1530s and late 1540s. Within a few years the subject of peasant festivals, which Beham had invented in Germany, moved to Flanders in the Netherlands where scatology played more minor roles and ultimately none at all by the time Bruegel designed his peasant festival paintings around 1567.

In attempting to understand the changing position of scatology, a number of explanations will be posited including: 1) the conservative effect the Council of Trent had on visual art of the second half of the sixteenth century, 2) the shift from the medium of prints to painting, 3) the civilizing process of Norbert Elias, and 4) the differences in German versus Flemish culture. Although no single explanation convincingly explains the cleansing of scatology from Bruegel's peasant festival imagery, the insistence on scatology in German art finds far easier explanation than its avoidance or elimination within Netherlandish art. Although further study is needed for a definitive answer to the question why Netherlandish art shied or turned away from scatology, I show here that the German emphasis on scatology is less a gloss added by the artist than a well-thought-out strategy that had deep cultural roots. My essay attempts to show that understanding a culture is important if we are to understand specific sixteenth-century visual imagery.

Peasant Festival Imagery from Dürer to Bruegel

During the 40 years from 1527 to 1567, when the first German and Netherlandish peasant festival images were made, the German peasant festival print with its emphasis on the body and scatology, in particular vomiting and defecating, was gradually transformed in the southern Netherlands into a cleaner image that made only discreet reference to the necessity of the body. By c. 1567, when Pieter Bruegel made his two well-known peasant festival paintings, no reference at all is made to scatology. Bruegel's *Peasant Wedding Celebration* and *Peasant Kermis* (Gibson, *Bruegel* 163; and Figure 10.1), both in the Kunsthistorisches Museum, Vienna, are scatologically free.

But in 1527 and 1528, the first two peasant festival images to include scatological reference were published at Nuremberg as woodcuts – works that constitute some of the earliest known visual renderings of the peasant festival theme.[2] Erhard Schön's design for his *Peasant Wedding Celebration* (Hollstein 48: 156), published by Hans Guldenmund who signed and dated the text 1527 at the lower right, and within the year Sebald Beham's *Kermis at Mögeldorf* (Hollstein 2:243), with small area designed by Schön,[3] were both designed by two men trained by Nuremberg's leading painter, Albrecht Dürer. These designers expanded their master's earlier interest in two or three peasants made between c. 1497 and 1514 (*Illustrated Bartsch* 10: 86, 89, 90) into bustling compositions filled with peasants dancing and drinking. Although Dürer's legacy can be noted generally in the approach to subject, composition, and forms, including the sturdy peasant figure, the inclusion of peasants vomiting and defecating introduces a new cultural aesthetic alien to the decorum of Dürer's work.

In the *Peasant Wedding Celebration*, Schön shows, at the right, peasants outdoors celebrating a peasant wedding with energetic dancing. At the left is the

Figure 10.1 Pieter Bruegel, *Peasant Kermis*, oil painting, c. 1567–68, Kunsthistorisches Museum, Vienna.

feast indoors where a number of men and the bride sit at a table and enjoy eating and drinking. The platters filled with roast fowl and sausages indicate the large quantities of food and drink consumed at the celebration. But it is the peasant man centrally placed at the bottom who directly points to the consequences of drink. He vomits onto the floor and a dog laps up what he emits. The tall drinking vessel on the table that looks much like a German *Weissbier* glass today indicates that an alcoholic beverage, perhaps beer, has caused the man's bodily distress. At the left, another peasant man seated on the same bench defecates in his pants, over the edge of the bench onto the floor.

The ideas shown here, although puzzling at first to the modern viewer, draw directly on the popular sixteenth-century notion of the four effects of wine, which Schön visualized in another woodcut from 1528 (Hollstein 47: 129), where the various reactions to wine drinking parallel sixteenth-century ideas on the four humors or temperaments. The consequences of drink for the sixteenth-century viewer were deemed comparable to the characteristics of the four animals whose blood provided the fertilizer for the grapes first grown by Noah: ape, lamb, lion or, here, bear, and pig – emblems of the four temperaments. Schön shows men and women sanguine as a lamb or as choleric as a bear, at the upper left and right, or as phlegmatic as a pig at the lower left and right. The ape at the lower right indicates the melancholic temperament.

Schön shows one man vomiting and two more defecating at the lower left, and together they are shown to act like pigs. Not coincidentally a four-legged pig laps up the drunkard's vomit and serves as an emblem for both swinish behavior in general and the phlegmatic temperament in particular. The accompanying text by Nuremberg's poet-cobbler, Hans Sachs, describes such a drinker under the third characteristic of wine, which Sachs explains includes overeating and drinking, drunkenness, farting, staggering, lying in filth, belching and farting like a pig, urinating in bed, and filthy language. Phlegmatics were believed to lose control of their bodily functions when drunk because of excess bodily fluids (Tlusty 54).

The concept of the effects of wine was also central to an understanding of the considerable body of printed images of kermis produced at Nuremberg between 1528 and 1535. Kermis was the most popular peasant festival in late medieval and Early Modern Germany and celebrated the anniversary of a church's consecration or its name saint. In 1528 Sebald Beham designed his *Kermis at Mögeldorf* woodcut, a wide, frieze-like image with four lines of text above. Beham's woodcut begins with a small building at left, probably a tavern (see Stewart, 'Taverns'), before which a pig consumes a mound of filth. To the right a table features men and women who eat and drink while shawm and bagpipe players provide music for fifteen dancing couples who fill the remainder of the long composition to the right.

The man seated on the bench, prominently placed at the beginning of the composition, vomits as the accompanying pig serves as an emblem of gluttonous behavior. The pig symbolized gluttony in German art during the first half of the sixteenth century as seen in Hans Burgkmair's woodcut from 1512 and in Georg Pencz's engraving from c. 1540 (Hollstein 5:98 and 31:212). In the *Kermis at Mögeldorf*, the viewer is informed that too much drinking and celebrating result in the kind of swinish behavior shown, vomiting. Here, the moralizing message and warning suggest that responses to such behavior could include revulsion and steering clear of the negative behavior shown. As we will see, other responses to such scatological imagery were possible as well.

Mögeldorf, the location featured in the title of Beham's kermis print, was a small village four miles east of Nuremberg's walls, still within its countryside, and less then a one-hour walk from Nuremberg's old town. Mögeldorf's kermises were so popular that half of Nuremberg's population are said to have visited them.[4] In 1528 Mögeldorf's population included only thirty peasants and forty-six rural workers, a small number overshadowed by Nuremberg's combined urban and rural populations estimated as high as eighty thousand inhabitants. We know that Mögeldorf's kermis festivities were celebrated at Pentecost in late May or early June and their attraction lay in their fortuitous location outside Nuremberg's one-mile city limit restricting brandy manufacture. As brandy increased in popularity as a wine mixer, so did Mögeldorf and its kermises.[5]

Beham's image of kermis showcasing a drunken, vomiting male peasant must have met with success and sold well despite, or because of, its emphasis on gluttony and emission judging from the several additional woodcut prints with the subject of kermis Beham had designed c. 1535, each showcasing a similar drunken peasant. In these woodcuts of kermis, Beham attempted a more narrative format that included additional kermis activities. In Beham's *Kermis (Erlangen)* (my 'Paper Festivals' figure 6; Hollstein 3:256 top), so-called because it is preserved in a unique impression housed in the University Library at Erlangen, Germany, Beham again includes peasants dancing to music beside an inn with a peasant vomiting before it. New here is the inclusion of a church in the background with a flag hanging from a window identifying the event as kermis (Pessler 2:109 and Kramer, *Bauern* 73). New too is the maypole with peasants dancing around it. Here the peasant man vomiting at the center is aided by a peasant woman who holds his head as if to comfort him, implying ample alcoholic drink. A dog samples the man's issue thereby serving the same function as the pig in the earlier peasant wedding celebration and *Kermis at Mögeldorf.*

Beham increased the number of scatological incidents to include defecation and a second example of vomiting. At the left, a man kneels on the ground behind the bench and vomits onto the ground. And behind the wattle fence above, to the left of the inn, another peasant man crouches as if defecating. We see only his head, in profile, but his short height suggests that he crouches or squats. In the right half of the composition (now lost but known through copies at Oxford and Gotha), Beham shows a man evacuating from behind in the midst of trees and bushes at far left, thereby switching the view of the defecator presented to the viewer. Here the man's buttocks, genitalia, and bowel movement are revealed. In the Oxford version (Figure 10.2), brown coloring added selectively throughout the print highlights the man's defecation, undoubtedly drawing attention to it.

Beham included two counts each of expulsion from top and bottom in his original kermis print at Erlangen also seen in the two copies at Oxford and Gotha. This emphasis on scatology points to a viewing audience that must have tolerated,

Figure 10.2 After Sebald Beham, *Kermis Oxford*, detail, Woodcut, after c. 1535, Ashmolean Museum.Oxford. Photo: author.

if not enjoyed, such bodily acts, although some viewers would have undoubtedly been offended by them, judging from the later printed comments of Walther Hermann Ryff, published 1548 in his *Vitruvius Teutsch*:

> Who can receive pleasure from a picture of a truly drunken peasant who shits and vomits behind a fence? Such repulsiveness [only] gives pleasure to someone who is improper, has the sense of a peasant, and who can inappropriately be called human. To the shame of painting, there are so many nowadays who draw and paint such inhuman things that ought to horrify a reasonable person. (Forssman ccxxxi^{v}; trans. mine).[6]

The vehemence of Ryff's reactions suggests that he would not have numbered among those buying, or even looking at, peasant festival prints. Buyers would have undoubtedly gone beyond tolerance to enjoyment, if not delight, of expulsion from top and bottom. The sheer number of peasant festival woodcuts produced in Nuremberg between 1528 and 1535 that include scatological incidents point less to tolerance alone than to enjoyment of scatology and to a variety of purposes that went well beyond moralizing and sinful associations.

Sebald Beham designed his largest and finest rendition of kermis in woodcut form, the *Large Kermis*, in 1535 for the Nuremberg publisher Albrecht Glockendon (Hollstein 3:255; Stewart, 'Paper Festivals' figure 9). Like the *Kermis (Erlangen)* and its two copies, the *Large Kermis* focuses on peasants eating and drinking, dancing, fighting, and playing games, but the arrangement of the scenes has been changed. Here Beham tones down the number of scatological incidents, including defecating, while underscoring the centrality of wine and vomiting, as in the *Kermis at Mögeldorf*. Beham also includes a mound of excrement to the left of the inn, below the wagon wheel, but he shows no one close by, thereby leaving the person or animal responsible open to the viewer. But human responsibility is showcased at center where Beham presents a man lying on a bench before the tavern (Stewart, 'Taverns' 104). The peasant man vomits a full stream onto the ground as a dog eagerly samples the fresh issue. Here Beham underscores the effects of wine by placing a trellis with grape leaves, vine, and a cluster of grapes above the drinker against the inn. Although the link with wine drinking is undoubtedly clear to the viewer, Beham links the emblem of vomiting man and dog with drunkenness in other ways as well.

The image of drunken man and dog visualize a contemporary proverb, the 'drunken matins' ('die truncken Metten'), which ironically refers to the vomit of a drunkard as singing in church. The expression itself can be found in both visual and textual form on the title page woodcut to a pamphlet entitled *A New Song. The Song is Called the Drunken Matins* (Stewart, 'Paper Festivals' figure 12), which was probably printed at Nuremberg during the 1530s. Sebastian Franck explains the expression in detail in his book of proverbs from 1541.[7] He writes that the drunken matins results from drinking so much that Bacchus throws the drinker under a bench, then the drinker begins to sing the drunken matins with long notes, a euphemism for vomiting, such that all dogs and pigs run to the drinker and gobble the song and the matins he produces. Franck calls such singing of the drunken matins debauchery.[8]

Beham's image of a drunken man in his *Large Kermis* appears to function on at least three levels: as the third effect of wine, as an emblem indicating the sin of gluttony, and as the visualization of a proverb. So understood, the long, thin loaf of bread placed over the drunkard's head, a curved, split loaf, ironically appears to stand in for a halo over the head of the vomiting man. Such apparently holier-than-thou moralizing views parallel those of Humanists and civic and local authorities in Germany who, during the early sixteenth century, believed that drinking was a serious social problem and, as elite members of society, vigorously railed against drinking in large quantities. Sebastian Franck summed up such attitudes when he stated that 'more people die from excessive eating and drinking than from the sword' ('Es sterben mehr leut von fressen vnnd sauffen, dann vom schwerd') (2: fol. 162).

But it is important to remember that a range of meanings could have been understood for images of individuals vomiting and defecating. People who drank to excess were nearly always men,[9] and in visual imagery and elsewhere anyone suffering from bodily expulsion could have been understood on several levels that mixed the kind of moralizing attitudes already seen with entertainment and delight. Although art history has been slow to understand scatology within historical contexts, historians of literature have studied the subject in some detail and their interpretations offer food for thought. Johannes Merkel (192–201) uses the term 'fecal comic' for the long tradition of delighting in scatology as a vehicle for the comic, through laughter and delight, in German carnival plays of the fifteenth and sixteenth centuries. Sidney Shrager (75–95) expands this understanding of the scatological for modern drama by calling attention to it by both joking and moralizing about it, a concept he terms 'scatological satire.'

Expanding on these ideas, recently Eckhard Bernstein (292) called attention to Barbara Correll's merging of these moralizing and pleasurable approaches by referring to the scatological antics of the German sixteenth-century figure Grobianus as the 'peepshow of civility;'[10] by this Correll (107) meant undesirable behaviors were both called attention to and delighted in as part of the civilizing process set forth by Elias. If we apply these various interpretations of scatology to the visual imagery of the sixteenth century, in particular to the German peasant festival woodcuts, our understanding of those images of peasants eliminating from top and bottom can similarly be understood in a variety of ways: as comic, satirical, and moralizing elements in service of the civilizing process.

Returning to Beham's kermis imagery, one notices that after 1535 Beham no longer provided kermis designs for Nuremberg publishers of woodcuts. Beham had moved west and north to Frankfurt am Main about 1531, and he returned to the subject of peasants celebrating in 1537, but this time in engraving not woodcut. Why Beham switched to engraving is unclear but there are several possibilities: perhaps he was dissatisfied with working with Nuremberg publishers over a long distance, or perhaps he received a one-time fee for each woodcut design and not for the number of impressions printed – the scant evidence from the period does suggest that sixteenth-century publishers, not the woodcut designers, sometimes owned and therefore kept the wood blocks. Assuming that Beham designed and cut his engraving plates, as earlier painter-engravers did, and perhaps printed his plates as well, perhaps he hoped to tap into a new and lucrative audience in his still relatively new home at Frankfurt.[11] During the decade from 1537 to 1547, Beham rethought his earlier large-scale woodcut peasant festival images and motifs and retooled them into small-scale engravings. Presumably such engravings were more expensive than woodcuts and enjoyed a more elite audience, assumptions that are generally true about woodcuts versus engravings, but may not necessarily be true for this particular case.

Beham's *Peasants' Feast* series, with twelve small engravings (c.2 ins x 1.5 ins) dated 1537 (Hollstein 3:94), includes what might be described as a little bit of something for everyone: musicians begin at the left, followed by eight couples dancing, the last male dancer vomiting, and ending with three engravings showing lovers and a man voiding. Partially disrobed lovers are caught by a woman who appears to be the man's wife, and the next engraving shows an older woman who wants to join in the love play. The last engraving is the scatologically centered scene, and perhaps the most scatologically rich of any we have seen. A squatting man vomits and simultaneously defecates, as a pig sniffs what he evacuates below. The male companion explains, 'You are really too shameless' ('Dv machst es gar zv grob'). He points to the man's bodily functions and calls attention to what he does and therefore what might be understood as his crude, inappropriate behavior. As Erasmus stated in 1530: 'It is impolite to greet someone who is urinating or defecating.... A well-bred person should always avoid exposing without necessity the parts to which nature has attached modesty....' (quoted in Elias 130). But the men Beham shows are peasants and undoubtedly not well bred. By showcasing behavior considered crude, perhaps peasant-like, Beham offers the viewer an image with a range of associations and meanings that can be described as entertaining, humorous, moralizing, and satirical.

Beham reused his series from 1537 a decade later in his *Peasants' Feast – Twelve Months* series dated 1546 and 1547 (Hollstein 3: 98). Using ten engravings, rather than the earlier twelve, Beham moved the musicians after his dancers, and concludes with three engravings centered around eating and drinking, fighting, and making love, including the effects of drink. In the last print (Figure 10.3; Bartsch 163), Beham combined two of his compositions from the series of 1537 into one. The lovers and the spectator at the left are reversed to mirror image and placed alongside a peasant man who appears to void simultaneously from top and bottom. The peasant lovers embrace eagerly, the man gropes beneath her skirt, as an older woman explains in the banderole above, 'I also want to' ('Ich wil avch mit'). In the right half of the print, which constitutes the last in the series, a squatting peasant vomits onto the ground while a pig, seated on the ground behind the man, appears to sniff the man's bowel movement, which is nearly invisible in most reproductions. The woven fence offers a modicum of privacy to the lovers in this scene, a fence that recalls Walther Ryff's question discussed above, 'Who can receive

Figure 10.3 Sebald Beham, *Peasants Behind a Fence*, engraving, 1547, no. 10 in the series *Peasant Festival – Twelve Months*, British Museum, London (courtesy Photographic Collection, Warburg Institute, University of London).

pleasure from a picture of a truly drunken peasant who shits and vomits behind a fence?

Judging from the existing numbers of woodcuts and engraved series of peasant festivals and celebrating peasants that Sebald Beham designed or inspired, not only did an audience exist for such images but the artist appears to have established a market for the subject between 1527 and 1547, first at Nuremberg and then Frankfurt am Main, and possibly beyond into the Netherlands. In 1549 the first of several Flemish prints opened up the topic that led, over the course of a decade, to the prints and paintings of Pieter Bruegel. Although art historians have generally distanced the Flemish examples from their German antecedents because of what they have called in the latter 'repugnant' or 'gross' behavior, the Flemish prints initially share such characteristics and only gradually, over the course of two decades, shed all scatological references.[12]

But the Flemish prints, then paintings, actually follow the German tradition in important ways. First, most of the images are prints which offered wide distribution and thus suggest a continuous broad audience that enjoyed and was willing to purchase the subject of peasant festival imagery on paper. Second, most of the Flemish images show peasants outdoors where they eat, drink sometimes to excess, dance, and enjoy games and other amusements. Rather than departing entirely from their German models, the Flemish examples continue in their footsteps in terms of the print technique and subjects shown.

The earliest Flemish prints of peasant festivals include peasants vomiting and defecating, yet they do so in smaller numbers and in more discreet locations than in Beham's prints. In this regard, we most definitely observe a move away from the scatological emphasis of the German prints. Over the course of less than twenty years, from the first Flemish print in 1549 to Pieter Bruegel's paintings – c. 1567–68 – the Flemish peasant festival images turned from a multi-peopled composition with a variety of energetic activities, including voiding from top and bottom, to monumental renderings of fewer, larger figures who comport themselves in more decorous manner. These peasants never vomit or defecate, at least where viewers can see, despite their drinking and fighting. And although they may dance with enormous codpieces displayed immediately in the viewer's face, as in Bruegel's painting of 1566 in The Detroit Institute of Arts (Gibson, *Bruegel* 161; Gibson, *Pieter Bruegel* figure 1), these peasants behave themselves with regard to bodily emissions. No vomiting or defecating is rendered visible.

The Flemish prints mark a move toward etching and Antwerp, two increasingly important aspects of the Northern European art world in the second half of the sixteenth century. The Flemish peasant festival prints employ etching on copper plates perhaps because etching was considerably faster than engraving on copper and reduced the time to cut a small plate, according to Landau and Parshall (30), from less than a week for a small engraving plate to less than a day for an etching. In the etching technique the design is incised by acid, rather than pushing a graver or burin into the plate. Thus the use of etching increased the ease and speed of producing images such as peasant festivals.[13] The Flemish festival prints also stress the Flemish port town of Antwerp, which rose to great power and importance during the course of the sixteenth century.

The Flemish prints mark the shift in the center of peasant festival imagery production away from Germany to the southern Netherlands and Flanders. Between 1553 and 1560, Pieter van der Borcht from Mechelen designed three etchings that weave the effects of drink and bodily elimination into the activities rendered; these etchings may have been printed by Bartholomeus de Momper at Antwerp. In van der Borcht's etching of 1553, the *Peasant Kermis* (Figure 10.4), for example, the scene teems with energy and tall, long-legged individuals, including a man who vomits at the lower right as an animal, possibly a dog, sniffs and expresses its interest.[14] In van der Borcht's later *Peasant Kermis* etching dated 1559 (Hollstein 3:467), a toddler appears to defecate, or urinate at center, from its bare bottom (Orenstein figures 54, 92) while a woman discreetly squats nearby to the left of the tree. Here only the child's voiding is shown directly; no examples of vomiting or defecating are included in the explicit manner of Beham's peasant festival prints despite van der Borcht's direct borrowing of the church procession from the former's *Large Kermis* woodcut (Stewart, 'Paper Festivals' figure 9; Hollstein 3: 255) pointing to cultural differences with regard to scatology.

By the time Pieter Bruegel's peasant festival designs were published as etching-engravings in Antwerp around 1559, scatological details were discreetly tucked into the compositions and inferred, rather than showcased.[15] In Bruegel's *Kermis at Hoboken* (Figure 10.5; Hollstein 3:208; Orenstein cat. 80), etched and engraved by Frans Hogenberg and published by Bartholomeus de Momper at Antwerp, Bruegel offers a panoramic view of the village of Hoboken outside Antwerp, where tavern and church provide the setting for dancing and drinking, archery and processions.[16] Two peasants appear to relieve themselves in the foreground: one man urinates before the tavern and another crouches behind the archery target at the left. Bruegel employs discretion for his incorporation of scatology by showing both men in back or side views; both views imply urination and a lower, side view defecation as well. These views suggest bodily acts without explicitly depicting them. A comparison with the descriptive, graphic back view employed by Beham is informative and again points to different cultural attitudes.

Bruegel emphasizes numbers of animals. He includes at least seven pigs and four birds in his print. These numbers indicate that the inclusion of the symbolic gluttonous animal *par excellence*, the pig, was a more appropriate form of representation for Bruegel and his Flemish audience than the overt depiction of swinish behavior itself. In his engraving of *Gluttony* or *Gula* from the Seven Deadly Sins series (Orenstein cat. 45), Bruegel showcases two pigs and two dogs in the foreground along with a man vomiting from a bridge, at lower left. Expelling from the top appears to have been acceptable here for depicting the sin of gluttony, but notable is the complete absence of the association between gluttonous behavior and defecation found in the German woodcuts and engravings.

It was only a small step from Bruegel's *Kermis at Hoboken* print to Bruegel's more familiar paintings on wood panel, the *Peasant Wedding Feast* and *Peasant Kermis* dating c. 1567–68 (Figure 10.1; and Gibson, *Bruegel* 163). Bruegel's

Figure 10.4 Pieter van der Borcht, *Peasant Kermis*, etching, 1553, Grafische Sammlung, Albertina, Vienna.

Figure 10.5 Pieter Bruegel and Bartholomaeus de Momper (publisher), *Kermis at Hoboken*, etching and engraving c. 1559, Rosenwald Collection, Photograph © 2002 Board of Trustees, National Gallery of Art, Washington.

audience was undoubtedly interested in a highly colored, unique original. Paintings on panel were considerably more expensive than woodcuts and engravings. Bruegel's turn toward panel painting for this subject may have also taken place at Antwerp where the Flemish etching-engravings were published. Although Bruegel resided in Brussels when he made these paintings toward the end of his life, he may have painted them for the Antwerp mint master, Jean Noirot, as Walter Gibson has recently shown (Gibson, 'Bruegel's Rustic Revels').

Bruegel's paintings of peasant festivals, often called festive peasants within art history, are simple in design and unusually large at approximately 4 feet high by 5 feet wide, compared to the German prints of the subject at 1 or 2 inches in height and width. In Bruegel's *Peasant Wedding Feast*, peasants feast simply indoors and drink moderately. In the *Peasant Kermis*, peasants dance energetically outdoors and engage in animated, even drunken discussion. Although Bruegel included a dog and tavern in each painting, bodily functions have been nearly eradicated from his visual vocabulary. In the *Peasant Kermis*, a standing man shown discreetly from behind faces a building and stands in a pose suggesting urination.

Except for this single detail, Bruegel's paintings mark the beginning of clean peasant festival imagery in Flemish art. Bruegel's peasant festival paintings may have enjoyed a more limited audience comprising merchants and Humanists as opposed to the broader audience enjoyed by prints.[17] If Bruegel's peasant festival paintings were in fact owned by a merchant, it is possible but by no means proven that such an owner sought paintings that were decorous, in keeping with the tastes of Flemish culture and classicizing Renaissance Humanism.

Explaining the Turn Away from Scatology

The turn away from the bodily element in Netherlandish paintings after the middle of the sixteenth century may be explained, at least in part, by several different but overlapping factors: by a different class of viewers who purchased paintings rather than prints, by the civilizing process explained by Elias, by a turn toward more conservative visual imagery after the mid-century as a result of the Council of Trent, and by differing German and Flemish cultural norms and tastes.

In the case of Bruegel's paintings, the omission of scatology might be first explained by the type of buyer who could afford to purchase such paintings. Panel paintings were more expensive because of their larger size and more expensive wood panel support than were printed works on paper, which were usually more modest in size due to the smaller size of individual sheets of paper then available. Bruegel's paintings could only have been afforded by a person of some wealth. As a rule, buyers of paintings were positioned higher in the social hierarchy than were buyers of prints. While prints were undoubtedly purchased by the wealthy and educated as well, they appear also to have been bought and viewed by the middle class, if not below. Because paper was less expensive than wooden panel, prints were produced in greater numbers, thereby reducing the cost of the individual work. In other words, prints enjoyed a more democratic audience.

The omission of scatology in the Flemish peasant festival images may have also been caused by an increasing sense of decorum in Renaissance society. Over the course of the sixteenth century, decorum increasingly made inroads into Northern society and influenced visual imagery in both Italy and the North as a result of the increasing influence of the classicizing aesthetic brought about by Humanism. Ryff's classicizing attitudes, discussed above, may serve as a good example here. In addition, after the mid-century a more censorious aesthetic resulted from the Catholic Counter-Reformation that directly influenced the kinds of images made and the kinds of changes made to existing ones. David Freedberg (*Power of Images*, 361–68, figures 165–76) has shown that, as a result of the Counter-Reformation, many prints with erotic subjects were censored in the late sixteenth century by reworking the plates dating from the first half of the century.

To understand the role the Counter-Reformation may have played in peasant festival imagery, let us investigate a particular peasant festival print by Beham. His *Nose Dance* woodcut was first published c. 1534 by Niklas Meldemann at Nuremberg and shows a group of mostly men dancing in a circle around a maypole. The print features large-nosed individuals who, according to beliefs at the time, would have possessed correspondingly large genitalia or breasts of corresponding size (Stewart, 'Large Noses' 346–47). By the late sixteenth century, the block was cut from its rectangular form with text below to a round form with decorative border (Stewart, 'Large Noses' figures 170, 176). In the altered form, the large size of the figures' noses was reduced in the block, and the exposed genitalia of the fool, at the lower right, were eliminated, thereby desexualizing the print. Beham's German woodcut appears to fit into the Counter-Reformation tendency toward prudery and censorship that Freedberg discusses with regard to Italy.

It was certainly nothing Catholic that had inspired the changes to the *Nose Dance* block, for there was no criticism of pope or emperor in the print. After the Council of Trent (1545–63) until the end of the century, increased efforts were made to revitalize Catholicism , not least by the bishops of Bamberg, who had jurisdiction over Nuremberg's religious affairs before it became Lutheran in 1525. In the last decade of the century Bishop Neithard von Thüngen (1591–98), a militant Counter-Reformationist, offered Nuremberg's small farmers the choice between emigration and conversion to Catholicism (Pfeiffer 267).

Given the Counter-Reformation pressures from beyond the city, it would be surprising if Nuremberg's publishers had not responded. Perhaps they felt the pinch under von Thüngen, or simply saw an opportunity to reissue prints and capitalize on their wood blocks. The last decade of the century, under von Thüngen, was contemporary with the reign of Clement VIII (1592–1605), whose objections to the nudes in Michelangelo's *Last Judgment* almost ended in the fresco's destruction. Just as 'a wind of pious and belligerent conservatism blew through the streets of Rome' in the 1560s after the Council of Trent (Landau and Parshall 361), a similar chill also seems to have blown through Nuremberg's streets under the influence of von Thüngen at the end of the sixteenth century.

The growing sense of decorum and conservatism throughout the sixteenth century may well, in itself, account for the general Flemish aversion to scatological images in peasant festival paintings and prints for such bodily functions are not decorous, but neither are they erotic and in need of being cleaned up. It is possible that scatological details were equally, if not more, objectionable because they showed aspects of human behavior that were increasingly becoming privatized, such as bathing, or that were deemed too private and indecorous to be shown, such as voiding. Elias's theory of a civilizing process offers another explanation why scatology had fallen into disfavor in Flanders by the second half of the century. Elias explained a process whereby society's elite and authorities encouraged social graces and manners across social classes, thereby pushing forward what Elias calls the changing 'shame frontier.' The definition of being 'civilized,' or 'uncivilized,' thus changed over time, as did attitudes toward the body. Elias includes sections dedicated to everyday behaviors relating to the body (which he calls natural functions) including blowing one's nose, spitting, and activities in the bedroom.

In 'Changes in Attitude Toward the Natural Functions,' Elias includes some remarkably poignant examples that aid our understanding of the civilizing process. As cited above, he quotes Erasmus of Rotterdam's *De civilitate morum puerilium*, or *Manners for Children*, published in 1530, and underscores attitudes about the increasing privatization of the body:

> It is impolite to greet someone who is urinating or defecating . . . .
>
> A well-bred person should always avoid exposing without necessity the parts to which nature has attached modesty. If necessity compels this, it should be done with decency and reserve, even if no witness is present. . . .
>
> To hold back urine is harmful to health, to pass it in secret betokens modesty. . . . The sound of farting, especially of those who stand on elevated ground, is horrible. One should make sacrifices with the buttocks firmly pressed together. (110)

As Elias indicates:

> The thoroughness, the extraordinary seriousness, and the complete freedom with which questions publicly discussed here that have subsequently become privatized to a high degree and overlain in social life with strong prohibitions shows particularly clearly the shift of the frontier of embarrassment.... (111)

Elias offers a few more examples from the sixteenth century that show both what people did and what they should not have done. Writing about what one should not do, but implying that people of the time did it (otherwise, why tell them *not* to do it?), della Casa writes in his *Galateo* of 1558:

> . . . it is not a refined habit, when coming across something disgusting in the street, as sometimes happens, to turn at once to one's companion and point it out to him.
>
> It is far less proper to hold out the stinking thing for the other to smell.... (quoted in Elias 111)

Elias also offers the Brunswick Court Regulations of 1589 which indicate that people defecated and urinated in inappropriate places: 'Let no one, whoever he may be, before, at, or after meals, early or late, foul the staircase, corridors, or closets with urine or other filth, but go to suitable, prescribed places for such relief' (quoted in Elias 111–12).

As these examples indicate, the body, its comportment, and its emissions became increasing concerns of society's elite during the sixteenth century in Germany and elsewhere in Europe. Erasmus, della Casa, and the Brunswick Court Regulations are typical of the elite interest in reshaping society and its members into a cleaner package, one that responded to elite expectations. By today's standards sixteenth-century Europe in general, as seen in these sources, and Nuremberg in particular were loud and dirty and the behavior there uncouth and violent. In Nuremberg people talked loudly in church and defecated on the street, even during visits by the emperor. One document stressed that residents should not use the streets in this manner during the emperor's visit to Nuremberg; rather, they should use the public *Sprachhäuser* or latrines.[18]

Edward Muir's recent discussion of the ambiguity of meaning held by the human body and the persistence of Christian ritual and the elite's attempt to privilege reason over passion offers another intersecting explanation for the persisting images of peasant festivals despite the objections and legislation against such real-life activities by both the secular and religious authorities and the elite. Although not mutually exclusive, these explanations parallel the changes seen in the German and Netherlandish images and, as such, they mark an important transition in Early Modern culture. And we should not forget, as discussed earlier, that the comic and satirical aspects of literature and carnival plays of the time offer entertaining antidotes to the overemphasis on the moralizing that persists in elite literature and documents from the time that have come down to us.

Although all these elements – comedy, satire, and the moralizing – worked both within the service of the civilizing process and independent of it, we should also not forget that sixteenth-century Germans enjoyed laughing at least as much as we do, with people undoubtedly laughing at different times and for different reasons. Carnival plays offer an instructive example for our discussion. In evaluating Mikhail Bakhtin on carnival, Edward Muir (91) discusses Bakhtin's concepts of the lower body, turning the world upside down, and defecation and sex, in a comic manner through 'grotesque realism' and the 'material lower bodily stratum.' The resulting festive laughter has, as two of its characteristics, the ambivalence and duality of the body. The former combines praise and abuse; the latter combines the lower bodily stratum (which includes defecation and perhaps vomiting) and the opposite, the upper stratum with pity and reason. This mixture of laughter and the bodily functions emphasizes, once again, that attitudes toward the body in the sixteenth century were numerous and nuanced.

No single explanation accounts for why images of scatology constituted such a visible part of the German peasant festival imagery or why their Flemish counterparts increasingly turned away from scatological elements to embrace the scatologically free paintings of Pieter Bruegel. Clearly the two cultures interpreted these forces differently. The interconnected nature of such cultural forces constitute the differences we must sometimes explain as culture, and in the end it may indeed have been German versus Flemish culture that embraced, or shunned imagery of the body in its visual art. Although it is tempting to argue that refined French court culture might be accountable, at least in part, for Flemish culture's more refined aesthetics, Per Nykrog's discussion (209–16) of the changing ways in which filth functioned in French literature from the late Middle Ages through the Baroque and neo-Classical periods, suggests otherwise, for he understood filth as an essential element for the civilizing process in France.

Further study is needed to wholly embrace cultural differences as the explanation why German peasant festival imagery is so scatologically charged in comparison to its somewhat later Flemish counterparts. Folklorist Alan Dundes rather compellingly argues that scatology, particularly defecation, constitutes an integral part of the German national character, and therefore German culture throughout the centuries. A small sample of his evidence includes (9–18) the frequent use of the word shit (*Scheiss*) in German popular expressions, a museum of chamber pots that opened in Munich in 1987 (22), and a pub in Berlin called the toilet or *Klo* (19). For the sixteenth century, the kermis woodcuts of Sebald Beham offer unquestionable evidence that scatology served both entertaining and didactic functions as extensions of German culture and its bawdy tastes at the time of the Reformation. These works also underscore Dundes's thesis that emphasizing shit is German in origin.

Notes

[1] Ideas for this essay were drawn from my earlier articles 'Paper Festivals,' 'Large Noses,' and 'Taverns.' Due to the increasing cost of publishing (photographs and permissions), I supplement the illustrations I include here with in-text references to published illustrations. I am grateful to Russell Ganim who generously translated French passages of Per Nykrog's text for me.

[2] Here 'peasant festivals' is meant to indicate representations of peasants at wedding celebrations and at kermis. Earlier Sebald Beham designed a *Spinning Bee* woodcut c. 1524, which falls outside this study. See Stewart 'Distaffs.'

[3] Sebald Beham appears to have designed the *Kermis at Mögeldorf* woodcut except for a small group of men and women at the beginning of the print, seated at the table at the back and right, designed by Erhard Schön. This kermis woodcut, along with several others made and published in Nuremberg, were earlier attributed to Barthel Beham after Heinrich Röttinger updated many of Gustav Pauli's attributions over a half century ago. I return the kermis woodcuts to the more convincing attribution of Pauli to Sebald Beham, Barthel's brother. I see no convincing evidence that Barthel Beham designed any woodcuts.

[4] Kindler text accompanying figure 66. See also Beyer.

[5] Braudel, *Structures* 241–43 on the appearance of brandy in Europe for the first time in the sixteenth century; he mentions that brandy was forbidden to be sold at Nuremberg on feast days in 1496. See Stewart 'Paper Festivals' 310 n.23

on Mögeldorf and the hordes that went to drink there, as mentioned by the town council in a decree of 1526.

[6] Ryff called himself Gualther Hermenius Rivius in the manner of the university educated. For more on Ryff, see Stewart, *Before Bruegel* Introduction.

[7] Earlier visual examples of the drunken matins theme illustrate books and pamphlets on the subject of drinking. I know of no discussion of the meaning of the proverb in printed form before Franck's of 1541.

[8] Franck part 2: fol. 148v. For Franck's quotation in the original German, see Stewart, 'Paper Festivals' 340 n.103.

[9] An exception to this nearly exclusively male equation with scatology is Daniel Hopfer's *Peasant Festival* etching datable c.1536 (Hollstein 15:108) where at least one woman vomits at the right. On women and drinking, see Tlusty 133–46. On drinking and masculine identity, see Tlusty 122–46 and Roper 145–67.

[10] Nykrog uses a similarly self-contradictory but titillating expression calling the use of ribaldry ('grivoiserie'), or baldly representing filth, as 'naive pornography' (209-16).

[11] Beham may have found engravings more profitable than woodcuts because he could reprint his engraving plates according to demand and thus profit directly from their popularity.

[12] For example, Moxey 39 and Gibson, *Pieter Bruegel* 14. No historical analysis was given; thus the twentieth-century distance from bodily functions appears to have obviated the necessity of serious study.

[13] Etchings may have offered an oppportunity for artists to experiment (see Altdorfer's etched landscapes from c. 1515–20) during the first half of the sixteenth century. While German artists struggled with the technical limitations of etching on iron, which rusted easily and suffered false biting, Netherlandish artists turned to etching on copper by 1520. See Hults 103–20 and Landau and Parshall 332–34.

[14] In the anonymous Flemish *Peasant Kermis* etching dated 1549, no scatological incidents are included. See Gibson, *Pieter Bruegel* 32.

[15] See the recent exhibition catalog of Bruegel's prints and drawings by Orenstein.

[16] For Bruegel's *Kermis at Hoboken* engraving from 1559 and the drawing for it, see also Freedberg, *Prints* cat. no. 28.

[17] For the more limited audience of merchants and Humanists for paintings, see Margaret Sullivan.

[18] The *Sprachhäuser* document is cited in full in Stewart, 'Paper Festivals' 304 n.6.

Chapter 11

Tamburlaine's Urine

Joseph Tate

Fever

This essay begins at an ending: Tamburlaine, the protagonist of Christopher Marlowe's late sixteenth-century, two-part historical tragedy *Tamburlaine the Great,* dies of a fever.[1] This is an unexpectedly mundane death given that he has been well established in parts 1 and 2 as the evil and unyielding 'scourge and wrath of God, / The only fear and terror of the world' (1.3.3.44–5).[2] That Tamburlaine – who with his men 'use to march upon the slaughtered foe, / Trampling their bowels with [their] horses' hooves' (1.3.2.149–50), and who would, 'without respect of sex, degree or age,' kill 'all his foes with fire and sword' (1.4.2.62–3), and whose men 'hoisted up' the 'slaughtered carcasses' (1.5.1.131) of virgins sacrificed to prevent his ravaging of Damascus – should die of a fever and not in battle is a narrative turn in itself that warrants closer attention. But a detail Marlowe appends to Tamburlaine's demise renders it all the more unusual: his death is heralded by a uroscopic diagnosis.

The primary sources at Marlowe's disposal show that historical fidelity accounts for why the character does not meet death on the battlefield. But these same sources are quiet on uroscopy, the conventional medical practice of rigorous and meticulous inspection of a patient's urine using the senses of sight, smell and taste. In 1883, the critics C.H. Herford and A. Wagner established two texts, one in Spanish and the other in Latin, as the principal sources for the play (Thomas and Tydeman 73). The texts are *Silva de Varia Lección* by Pedro Mexía, printed in Seville in 1541, and *Magni Tamerlanis Scytharum Imperatoris Vita* by Petrus Perondinus, a Latin text published in Florence in 1553 which drew on Mexía's work. While Marlowe may have read either of these books in their respective languages, the Spanish text was available in two English adaptations circulating before the date ascribed to the play: Sir Thomas Fortescue's *The Foreste or Collection of Histories* published in 1571, and George Whetstone's *The English Myrror* of 1586 (Thomas and Tydeman 74). Though none of the play's principal sources, their later adaptations, or the twelve other possibilities cataloged in Vivien Thomas and William Tydeman's 1994 study of Marlowe's source materials, has Tamburlaine die on the battlefield, neither does any source particularize the circumstances of Tamburlaine's last moments.

His death, in fact, receives little attention at all. One medieval account uncovered by Thomas and Tydeman claims that Tamburlaine died in his sleep, 'through choking in his own bedchamber while his guard slept' (quoted in Thomas and Tydeman 121). The later English versions of the story are even less specific. Whetstone noted with no elaboration that Tamburlaine 'by the course of nature died' (82), and Thomas Fortescue plainly stated that, 'as a man in the ende, he paieth, the debte due vnto nature' (87). These sparse details have prompted equally sparse commentary. As Johnstone Parr argues, Marlowe, 'did not get specifically from his source material the cause and the manner of Tamburlaine's death' (5). Likewise Tatiana Wolff, in her edition of the play, concurs with Una Ellis-Fermor in proclaiming: 'Marlowe had no precedent for the death scene of Tamburlaine' (267).[3] Had 253 lines of verse not been given to what other authors relegated to an insubstantial subordinate clause, an inquiry could satisfactorily end here. Yet to further complicate matters, the history of Tamburlaine, the fourteenth-century Tartar emperor, was well known to Early Modern audiences before it was dramatized by Marlowe. Thomas and Tydeman cite and confirm the critic W.L. Godshalk's contention that the Tamburlaine story, 'may be found in as many as one hundred Renaissance sources, told from varying points of view' (quoted in Thomas and Tydeman 70). If there existed no precedent in the most immediate sources, from what materials did Marlowe construct details for a tale well known long before being brought to the English stage?[4] An answer, I argue, is in Tamburlaine's urine.

Diagnosis

In the final scene of part 2, Tamburlaine arrives accompanied by Amyras and Celebinus, his two living sons (having earlier killed his third son Calyphas for 'folly, sloth, and damnèd idleness' [2.4.1.126]), astride a chariot drawn by the conquered kings of Natolia and Jerusalem.[5] He is attended also by a small cadre of physicians. Once on the stage, he commands the armies awaiting his arrival to wage unending war with the deities that disturb his health:

> Shall sickness prove me now to be a man
> That have been termed the terror of the world?
> Techelles and the rest, come take your swords
> And threaten him whose hand afflicts my soul;
> Come, let us march against the powers of heaven,
> And set black streamers in the firmament,
> To signify the slaughter of the gods.
> Ah friends, what shall I do? I cannot stand.
> Come, carry me to war against the gods. (2.5.3.44–52)

At this, Tamburlaine's high-ranking officers caution against such 'impatient words,' which they maintain add 'much danger' to his malady (2.5.3.54–5). For his own good, his follower Techelles chides him: 'Sit still' (2.5.3.64); and, unknowingly foreshadowing Tamburlaine's end, tells his leader that the sickness 'cannot last, it is so violent' (2.5.3.65). Having been so violent himself, Tamburlaine, like his fever, will not last.

Despite the expressed concern of Techelles and others around him, Tamburlaine's verbal fury refuses to abate until he is interrupted by a physician's instruction to drink a restorative potion that will soothe his restless spirits. The patient takes this opportunity to ask the doctor: 'Tell me what think you of my sickness now?' (2.5.3.81). The physician responds with nothing less than bad news: 'I view'd your urine, and the hypostasis, / Thick and obscure, doth make your danger great' (2.5.3.82–3). The diagnosis continues with an account of Tamburlaine's humoral condition using uroscopy.[6] Well before the fourteenth-century reign of Timur, the Tartar emperor and historical model for the fictional Tamburlaine, Judeo-Christian and Islamic doctors alike considered uroscopy their primary diagnostic tool, a tradition supported by the venerated Hippocratic corpus. Nearly two thousand years after Hippocrates, when Marlowe was composing *Tamburlaine*, uroscopy was still widely considered a reliable method.[7] Using the eyes, nose and tongue, Tamburlaine's physician would have been able to identify scores, perhaps hundreds, of ailments in a single glass vial of urine.

The physician diagnoses Tamburlaine's particular ailment as follows:

> Your veins are full of accidental heat
> Whereby the moisture of your blood is dried:
> The humidum and calor, which some hold
> Is not a parcel of the elements
> But of a substance more divine and pure,
> Is almost clean extinguishèd and spent;
> Which, being the cause of life, imports your death. (2.5.3.84–90)

Tamburlaine's 'humidum,' or radical moisture, and his 'calor,' or natural heat, are both failing. Together, this moisture and heat was considered the generator of the vital spirits which mediated all physiological activity.[8] However, the accidental, or excessive, heat has dried the blood, leaving too little natural heat to move the spirits.[9] But lack of moisture is the central problem:

> Besides, my lord, this day is critical,
> Dangerous to those whose crisis is as yours:
> Your artiers, which alongst the veins convey
> The lively spirits which the heart engenders,
> Are parched and void of spirit, that the soul,
> Wanting those organons by which it moves
> Cannot endure, by argument of art. (2.5.3.91–7)[10]

Wolff concisely clarifies the condition in her edition's notes for *Tamburlaine*: 'The arteries should convey to the veins the blood that the heart produces; but they are dry and empty and so the soul is deprived of the spirit which feeds and moves it and is about to die' (267). Rephrased, Tamburlaine's intense passions, probably choleric, have overheated and dried the vital moisture required by the body's spirits, and this leads to death. Further, abnormally thick and dark urine or, as the physician calls it, 'thick and obscure' urine, is symptomatic of a severe fever or a febrile distemper.[11] His choler which, according to humoral theory is hot and dry, has fevered and dehydrated him.[12]

The physician's diagnosis in the play shows that Marlowe was more than somewhat familiar with contemporary uroscopic procedures. As a student at Cambridge, this information likely came to him via the standard Galenic texts that were faithful revisions of the Hippocratic discipline. The physician's diagnosis of Tamburlaine's urine resonates more widely with the play once it is contextualized amid the uroscopic manuals circulating in the period, all of which imply a peculiarly powerful valence to urine that remains uniquely stable across the period's competing medical ideologies.[13] Therefore, in what follows I turn attention to one possible model for Tamburlaine's physician's methods as they appear in the comprehensive uroscopic manual by Robert Recorde, *The Vrinal of Physick*, first published in 1547, and then numerous times subsequently under varying titles.[14] As Shakespearean scholar Gail Kern Paster writes, one finds that 'the categorical complexity of traditional uroscopy is likely to amaze the modern reader' (41). The complexity of Recorde's text is no less amazing.

Tamburlaine's physician would have reached his conclusions by examining a glass vial of the patient's urine following methods not unlike those set down by Recorde. To begin, Recorde asserts, in line with other writers from the period, that urine is a waste substance of blood that separates, 'as whey from mylke' (11^v). Sir Thomas Elyot, a contemporary of Recorde's, wrote in the *Castel of Helthe*, a treatise on maintaining everyday health, that 'Vrine is the watry substance of the bloudde, lyke as whay is of mylke' (53). Years later on the continent, René Descartes also has urine resulting from blood in his *Treatise on Man*: 'There are also some [parts of the blood] that flow out as urine through the flesh of the kidneys, and as sweat or other excrements through the skin' (104). Urine, considered the excrement of blood, attained status as a dependable gauge of internal physiological processes, and investigation of a patient's urine began with the proper urinal.

A urinal, also called a 'jordan' after the river Jordan of Christian significance, was a bladder-shaped glass container with a long, uncertain history. More specifically, as Recorde describes it, the urinal

> shulde be of pure clere glasse, not thyck, nor greene in colour, without blottes or spots in it, not flat at the bottom, nor too wyde in the necke, but widest in thc myddell, and narrow styll towarde both endes, like the facyon commenly of an egg, or of a very bladder beyng mesurably blowen (for the Vrinall should represent the bladder of a man) and so shall every thyng be sene in his dew place and coloure. (14^v)

Figure 11.1 shows what is presumably Recorde's illustration of a human bladder, labeled F. This compares to Figure 11.2, a urinal with a bladder-like shape that has the due place to which Recorde refers clearly marked. As the image shows, a

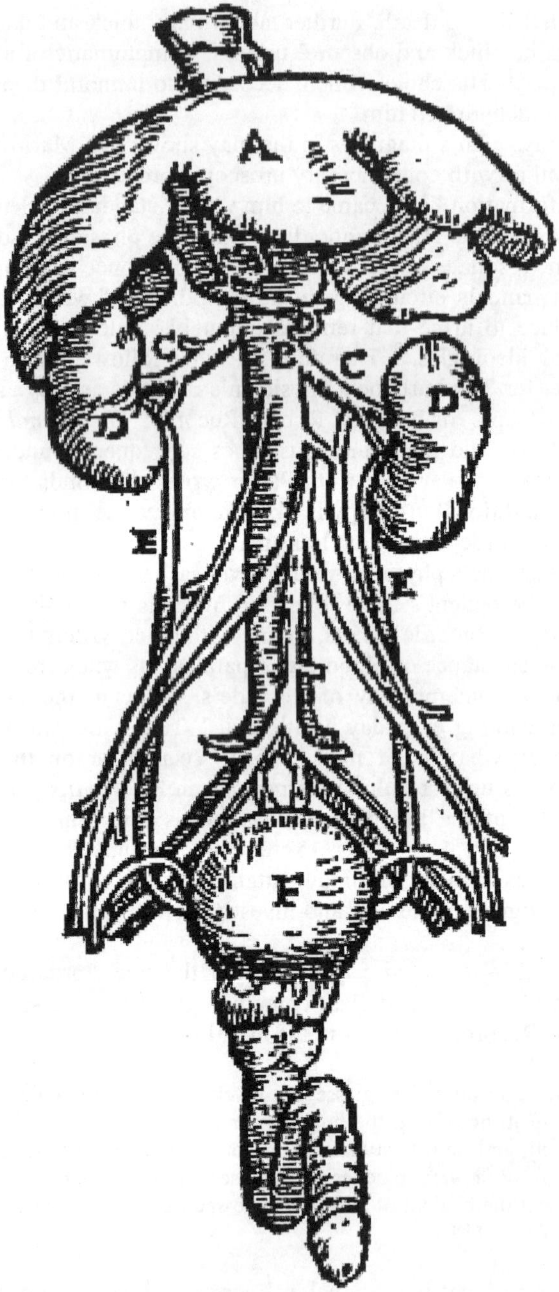

Figure 11.1 Robert Recorde, *The Vrinal of Physick* 13[r]. From the collection of the Royal College of Physicians of London.

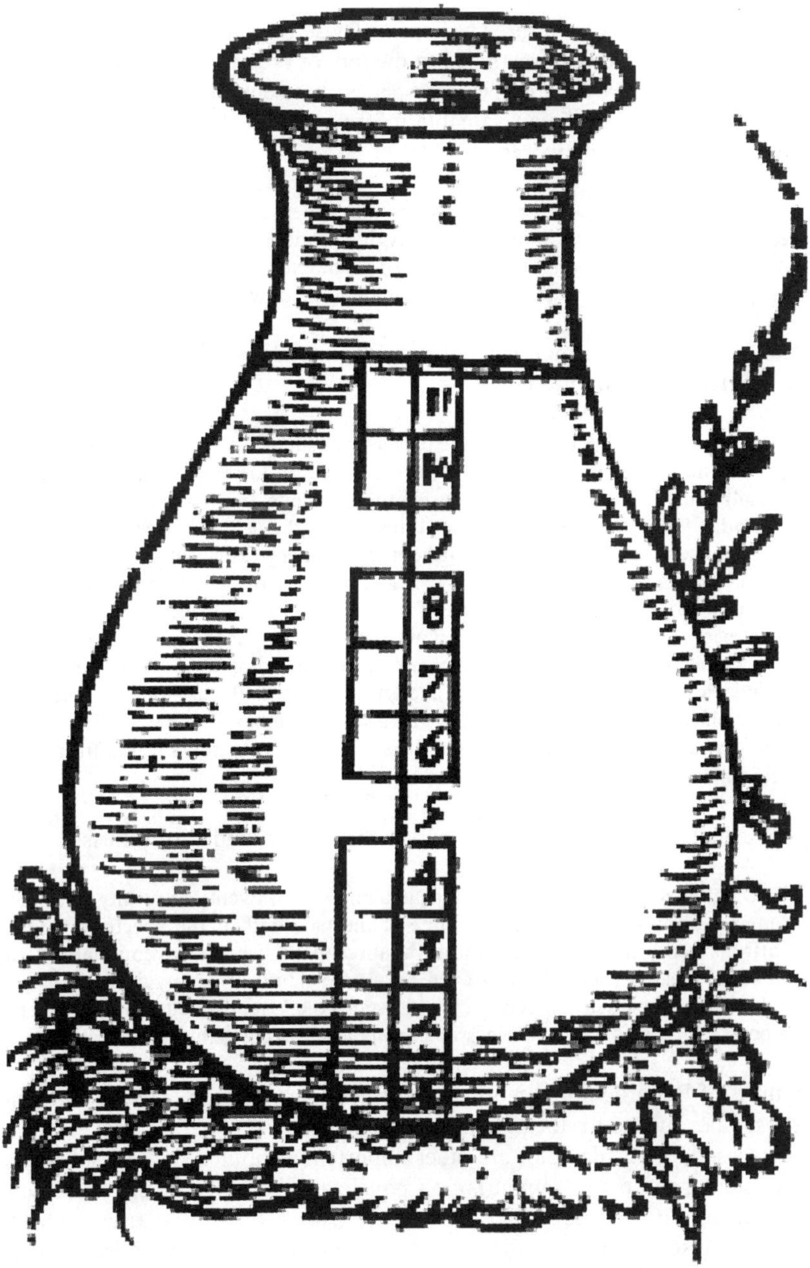

Figure 11.2 Robert Recorde, *The Vrinal of Physick* 18^r. From the collection of the Royal College of Physicians of London.

patient's urine was divided into three regions.[15] These regions were then subivided into 11 parts, with 1 at the bottom and 11 at the top, or crown, as it was called; levels 5 and 9 were 'the voyde roume,' or spaces, between the regions (Recorde 17^{v}).Each level told the physician valuable information on the corresponding part or parts of the body. Recorde writes:

> ...you shal obserue a certayne proportion that is betweene the partys of the vyrne, & the partys of mannys body. the [*sic*] hyghest parte of the vyrne doth betokyn the highest parte of the body, namely the hed ... the nether regyon of the vryne doth purporte the lowest partys of man, from the bowels downe ward. (51^{v})

Getting the urine into the urinal was not a task to be taken lightly. Recorde gives the following detailed instructions:

> ... you must take heede (whether the pacient be man or woman) that they make not their urine in a nother vessell fyrst (as many use to do) & then powre it into the urinall when it is setled, for that causeth much deceite & error in the judgement of it. And if that the pacyent can not well make it in the urinall, other by wekenesse, or any other cause, then let them make it in an other vessel, but se that it be clene & and dry: & as sone as the water is made, powre it furth straight wayes into the urinall all together, & leve no parte of it out, as sum curious folk do use to put the clere part only into the urinall & cast away the dregges: as though it stode not with their honesty to brynge soche foule gere to the Physicion, other any lyke foolysh mynd. Powre it therfore in wholly, and let not the urinall stand open, namely in a dusty place, but stopp it close with a glove or other lether, & not with cloth, paper, nor hey, and let it be brought to the Physicion within vi. hours at the furdest, for after that tyme it can not well be judged. (14^{r-v})

Once in the urinal, color was then examined as the primary index of illness:

> To the trained uroscopist, urines came in a rainbow of twenty or twenty-one colors ranging from several shades of white at the pale end of the spectrum through saffrons, reds, greens, blacks. These variations gave rise to conventional descriptions of great particularity and even beauty. We find urine 'nygh as yelowe as saffron of the garden' or 'reed as a brennynge cole,' 'redde as it were the lyver of a beast,' 'grene as wortes,' 'black & shinyng as a Ravens fether.' (Paster 41–2)

Nearly a third of Recorde's text is given to urine's vast color spectrum, and what each of those hues means for the patient concerned.

Assessed alongside color are other important attributes such as substance, quantity, contents, and taste. Tamburlaine's physician does not take special note of quantity or taste, but he does reveal that the hypostasis, or sediment, is thick and dark. Concerning sediment, Recorde warns that the worst colors are 'blew, ash colour and blacke ... and the nearer they cleue to the bottom of vrinal, the worse they are' (42^{v}). Generally, he says, black sediment comes from either 'excessyue colde other els excessyue heate' (42^{v}). Early in the text, Recorde equivocates on the danger of thick black urine:

> … if blacke coloure appere in thicke vryne, it betokeneth somtyme well, as in the ende of the feuer quarten, and of melancolyke madnesse / for it betokeneth that the melancolyke mater, which caused the diseases, bothe auoyde out. But sumtymes is an euyl token, for it sygnifyeth that either the blood is burned through excedig heat, or els that natural heate is clene quenched through dedly colde, and therefore is commonly called a dedly sign, namely in sharpe Agues, if it have an euyll sauour. And so meneth Galen, when he saith, that he markyd: the thicker that a blacke water is, the worse it is: and more ouer: that he neuer saw any escape, which made soch vrine. ($30^{r–v}$)

Recorde's 'sumtymes' is not encouraging, however. The 'sumtymes' when black urine is not lethal is on certain astrologically marked days: 'in moderate Agues, if soch blacke vryne do appere, it is a token of death, excepte it be on sum cryticall dayes' (36^{v}).

Later, all equivocation disappears as if doubt never existed:

> Now as for blacke, I nede not to speake any whyt: for as all men doo know hit, so these letters do shew it / which though of all other hit be the most dedlye, yet is hit surely of all the most myghty: for hit ouer cumyth all coloures, & none can chaunge hyt: so that well yt may be called the colour of death. For as death ouer-cummyth all bodyes, so blacke doth dampne all colours: besyde that it is the messenger & and token of death, which is the ende of all thiges, & and blacke the end of colours. ($68^{r–v}$)

The physician's diagnosis of Tamburlaine's urine is accurate, or at least in keeping with Early Modern practice, if we follow Recorde's manual; the danger, as he claims, is great.[16] Galen concurs with what we find in Hippocrates, who claimed that, 'in adult men and women the black is of all kinds of urine the worst' (48–9).

For twenty-first-century urology, brown or black urine still represents 'serious pathologic conditions that require prompt diagnosis and treatment' (Ringsrud and Linné 37). Modern urinalysis concludes that Tamburlaine's condition is probably nephritis, or inflammation of the kidneys, a fitting assessment given that the kidneys, known as the 'reins,' were considered the seat of the passions in Marlowe's time.[17] Since Tamburlaine is experiencing an excess of choler, whose corresponding element is fire, it seems no surprise that modern medicine would also determine that his kidneys are inflamed – that is, on fire. Black urine, whatever it might culturally signify now or then, indicates a physically severe condition. As Recorde claims, with support from Galen and Hippocrates, no one escapes black urine, and neither does Tamburlaine; within minutes of receiving the diagnosis, he dies.

Notably, Tamburlaine's fatal disease corresponds descriptively with Tamburlaine's lethality. Compare uroscopist John Fletcher's comment that 'Blacke vrines are mortall in all ages and persons' (44), to the description quoted above of Tamburlaine who kills, 'Without respect of sex, degree or age' (1.4.2.62); both black urine and Tamburlaine himself kill indiscriminately. In this way, recourse to

Early Modern uroscopy helps locate a correspondence between Tamburlaine's physical illness and his violent, transgressive character – a correspondence that foregrounds the play's larger concerns with geographic and biological boundary making and breaking.

Boundaries

That geographic boundaries are a focus of the play is well established in Ethel Seaton's oft-cited 1924 study entitled 'Marlowe's Map,' wherein the author persuasively argues the extent to which the play is indebted to Abraham Ortelius's *Theatrum Orbis Terrarum*. Recent critic David H. Thurn has stated the matter plainly: 'Marlowe composed *Tamburlaine* with a copy of Ortelius' *Theatrum Orbis Terrarum* on his desk' (180).[18] As if to establish this thematic insistence, act 1, scene 1 of part 2 begins at the Classical geographic divider between east and west, 'on Danubius's banks' (2.1.1.7). During the Early Modern period and centuries before, the Danube acted as a geographical feature that separated east from west, orient from occident. The modern geographer Paul Robert Magocsi writes:

> For the first four centuries of the common era, East Central Europe was divided into two spheres: the civilized world of the Roman Empire, and the uncivilized expanse of sparsely settled lands inhabited by sedentary peoples such as the Germanic Goths, Vandals, and Sueves, and the Asiatic Alans and Huns. The basic dividing line between these two spheres was in large measure formed by the Rhine and Danube rivers, along which the Roman Empire established its northern frontier known as the limes. (5)

For the Roman Empire, the Danube formed its northernmost limit beyond which dwelled barbarity. Early Modern texts confirm the Danube's status as the primary, though not always unquestioned, separating line between familiar self and foreign other, Europe and Asia. The 1606 London edition of Ortelius's *Theatrum* registers the problem as follows: 'by what limits they [Asia and Europe] are there distinguished, the ancient and the later writers do not altogether agree,' yet 'that *Donaw* (Danubius) doth diuide Asia from *Europe*, Seneca in the sixth booke *Natural*, doth manifestly affirme' (259).

Throughout the history of Eastern Central Europe and Europe at large, the Danube has been a problematic geographical and socio-cultural feature acting both as container and divider of peoples. Rivers generally still resist categorization. Returning to Magocsi, we find that present-day geographers deny that:

> … rivers can serve as borders of geographical units, because by their very nature rivers are bodies of water that unify rather than divide surrounding areas. This may be true in purely geographical terms, but it equally true that rivers have been used as boundaries of political units, thus often deliberately restricting or even eliminating their otherwise 'natural' unifying characteristics. (2)

The scene confirms this reading. Orcanes, the king of Natolia, stands on the eastern bank gazing at the king of Hungary's forces and wonders aloud to his ally Gazellus, the Viceroy of Byron, whether they should 'parley with the Christian / Or cross the stream and meet him in the field?' (2.1.1.11–12). Gazellus, knowing and recognizing limits, recommends they 'treat of peace' (2.1.1.14) and not cross the stream as they have more pressing matters: Tamburlaine's imminent arrival.

The Danube is not a trivial setting for the Natolian Muslims and Christian Hungarians to meet to do battle, if only because five more times in fewer than 150 lines we are reminded of the location: 'Two hundred leagues, and on Danubius' banks' (2.1.1.7); 'Danubius' stream, that runs to Trebizond,' (2.1.1.33); 'The Terrene main, wherein Danubius falls' (2.1.1.37); 'We with our peers have crossed Danubius' stream' (2.1.1.79); 'Confirmed this league beyond Danubius' stream' (2.1.1.149). However, although the division between east and west is highlighted throughout, the argument that the play examines Orientalism as we now know it is not sustainable. As Jonathan Burton has argued persuasively in relation to *Tamburlaine*, 'the conditions for Orientalism did not exist in the England of the sixteenth century. Orientalism sprang from a specific conjunction of socioeconomic and political circumstances that had not yet occurred in the Early Modern period' (152). The particularity of Tamburlaine's otherness is not implicated in the play's exploration of boundary transgression as much as his unmitigated desire to transgress.

The play's thematization of transgression has led critics to read the play as an ultimately liberating narrative. Representatively, Jonathan Dollimore argues in his foundational study *Radical Tragedy* that 'Tamburlaine's will to power leads to liberation through transgression' (114).[19] Tamburlaine's aspirations are indeed transgressive, but their liberating status is worth investigating. Tamburlaine's particular transgressions are redrawn as Bakhtinian inversions by Mark Thornton Burnett in his essay '*Tamburlaine* and the Body,' a frequently cited contemporary treatment of the play's corporeal concerns. The transgressive inversions in the play are numerous. As Peter Stallybrass notes, for example, the the image of the foot throughout *Tamburlaine* is a primary inscriber of power ('Footnotes' 314). Likewise, Burnett writes that Tamburlaine ensures time and again, that:

> … his enemies are brought low, reduced to the grotesque. A series of 'uncrownings' in the play, to use Bakhtin's term, shows this impulse meeting with considerable success. Of the executionary methods available to Tamburlaine as an Elizabethan tyrant, his preferred is the act of penetration: he punctures, perforates, dismembers, breaches, cleaves bowels, lances, spears and brings opened foes to his feet. (*Tamburlaine* 35)

In this reading, the uroscopic diagnosis that accompanies Tamburlaine's death helps situate the play within a carnivalesque tradition given that, 'scenes representing the physician at the patient's bedside were a recurrent form of scurrilous forms of "grotesque" popular literature' (Burnett, *Tamburlaine* 42).[20] Thus both the physical violence Tamburlaine inflicts and the biological illness he suffers is a form of 'grotesque realism' (Bakhtin, *Rabelais* 370) whereby inversion becomes the hermeneutic key:

> Down, inside out, vice versa, upside down, such is the direction of all these movements. All of them thrust down, turn over, push headfirst, transfer top to bottom, and bottom to top, both in the literal sense of space, and in the metaphorical meaning of the image. (370)

Violence, especially, becomes grotesque, not abject: 'We also see the downward movement in fights, beatings, and blows; they throw the adversary to the ground, trample him into the earth. They bury their victim. But at the same time they are creative; they sow and harvest' (370). That Tamburlaine's trampling of bowels (1.3.2.149–50) is at all regenerative is questionable, however.

While I agree with Burnett that 'in a scatological vein, the play delineates an upstart, a hero from the margins who throws into disarray established orthodoxies and who interrogates boundary structures, dividing lines and the mechanisms employed to institute social discipline' (*Tamburlaine* 43), I hesitate to concur with Dollimore's and Burnett's parallel assumption that Tamburlaine's particular form of violence is finally emancipatory. While *Tamburlaine* renders, in narrative form, the inversion of hierarchies in a quasi-carnivalesque fashion (e.g., Tamburlaine himself is a low-born shepherd who eventually rules much of the Middle East and portions of Eastern Europe), the tale does not evoke the emancipatory laughter that Bakhtin found in such inversions. Far from being a form of grotesque, subversive comedy, the narrative of the 'Bloody and insatiate Tamburlaine' (1.2.7.11) gives a quintessential example of abjection. As Julia Kristeva defines it: 'Abjection ... is immoral, sinister, scheming, and shady: a terror that dissembles, a hatred that smiles, a passion that uses the body for barter instead of inflaming it' (4). Kristeva's theory of abjection recuperates the play's concrete evocation of utterly laughter-less horror, especially as Kristeva's work owes much to Mary Douglas's *Purity and Danger*.[21] Significantly for this essay, Douglas discussed the boundaries or margins of the body as being especially invested with cultural significance: 'all margins are dangerous,' and the bodily relict of urine, like other excrement or bodily fluids, has 'traversed the boundary of the body' (122). To use Douglas's words, in this play the 'symbolism of the body's boundaries is used in this kind of unfunny wit to express danger to community boundaries' (124). Urine, Kristeva confirms, 'signifies the other side of the border, the place where I am not' (3) – that is, the place where Tamburlaine eventually arrives: a 'convocation of politic worms,' as Hamlet would say (4.3). Urine, therefore, and especially in this instance, is unfunny.

Treating Marlowe's narrative as an instance of the carnivalesque, I hold, requires one to forsake an essential element of Bakhtin's theory: his appraisal of laughter as liberation, especially as laughter pertains to matters of the grotesque lower bodily stratum (*Rabelais* 376).[22] Attention to the material, lower body, Bakhtin argues, 'gaily and simultaneously materializes and unburdens' (376). But such attention, I contend, cannot do both. The material has weight, physically and metaphorically, and laughter potentially forsakes this relation. Laughter renders the lower body 'more material, closer to man and his body, more understandable, and lighter in the bodily sense' (*Rabelais* 380). The body is approached through laughter that lightens. Bakhtin's almost unnoticeable figurative transformation by which the body is lightened renders it safe. The darkness, the evil of fecal matter is lightened, made lighter, in color and tone. Through Kristeva via Douglas, the body's materiality can be retained without lightening it, without turning the scatological into 'a gay and free play with objects' (Bakhtin, *Rabelais* 380). The mortal boundary that urine marks in the final scene thus retains its gravity. Urine as a boundary for Tamburlaine is not figuratively light, it is literally dark – thick and obscure – and is not laughable. The black streamers he would set in the firmament above appear instead below, in the sediment of his urine.

The urine, thick and obscure, is the evidence of Tamburlaine's impending mortality; the one boundary he cannot cross is his own. Throughout the play, however, we witness Tamburlaine's consistent refusal to acknowledge any sort of boundary – abstract, physical, or otherwise. Geographically and politically, Tamburlaine extends the margins of his power with terrible precision: in both their ideational and physical forms, numerous nations and their body politics collapse while mutilated, lifeless bodies are discarded like so much untouchable waste.

Human bodies are reduced to disposable objects within a matrix of subjugation where a certain, if arbitrarily determined, subject (i.e., Tamburlaine and his martially-trained subjects) exercises an absolute agency. Living bodies become exuviae, shed with sickening expedience. This perverted transubstantiation, this translation of animated body into inanimate refuse, is, for Tamburlaine, sublime. He carouses before the devastation, reveling in what Kristeva identifies as an 'ecstasy before the corpse,' an 'ecstasy before death' (150), an ecstasy that is nothing less than abject; his joy summons our feelings of abjection. Laughter in these moments would be avoidance of the obvious devastation. As the philosopher Henri Bergson has argued, laughter 'demands something like a momentary anesthesia of the heart' (11), an anesthetic distance that the play's action arguably denies.

In the play's closing scene, however, Tamburlaine's visceral ecstasy of sublimity before the cadaver fails (it falls, *cadere*) when confronted with the essential artifacts of the body, the excrement that falls from life.[23] Excrement is here defined as: 'These bodily fluids, this defilement, this shit are what life withstands, hardly and with difficulty, on the part of death' (Kristeva 3). The defilement that life withstands cannot, at this moment, be withstood. Defilement, Kristeva argues, 'is what is jettisoned from the "*symbolic system*"' (65), and the physician, perhaps only in figurative language, but still to powerful effect, reinjects defilement: he brings Tamburlaine his urine, an indisputable fact of his existence and the truth of its end.

After the uroscopic diagnosis is presented, once his limits have been established, Tamburlaine appropriately rehearses the borders he disestablished, that he traversed; he soliloquizes on the slow, returning sweep from Scythia to Persia, Armenia, Egypt, Arabia, Alexandria, India, Nubia, Zanzibar, Greece, through Asia and Africa, and back again to Scythia. The catalog of dissolved boundaries and dispersed peoples becomes, however, the definitive and terminal demarcation of his reign. With the map before him, he laments the unknown western lands he never will traverse, the new world he never will subdue. Close to tears, he asks rhetorically, but with all sincerity, 'shall I die, and this unconquered?' (2.5.3.151).[24] As a horde of opposing armies, preparing to regain the borders they had lost, waits angrily within sight of Tamburlaine's military encampment, the answer to his question stands thick and obscure in the vial of his urine: a bleak and terminal 'yes.'

Notes

[1] The play was first published in 1590, but its date of composition and performance is widely supposed to have been 1587. J.S. Cunningham's introduction to his 1981 edition of *Tamburlaine* discusses the play's textual and performance history in useful detail.

[2] All *Tamburlaine* quotes are from Cunningham's edition.

[3] Critic Peter Berek writes, 'Invulnerable heroes are hard to incorporate into plots, and Marlowe notoriously had trouble finding a satisfactory way to end *Tamburlaine* Part II' (39). Though Berek deems Marlowe's trouble notorious, no amount of research has turned up the writer's block in question.

[4] My assumption, a dangerous one perhaps, is that Marlowe's play does not give to airy nothing a local habitation and a name. Though Harry Levin maintains that *Tamburlaine* 'exhausted Marlowe's source material' and that 'when the play's unprecedented success demanded a sequel, he was thrown back upon his own imaginative devices' (35), Levin seems unaware that the so-called imaginative devices are simply more source materials. He notes himself that Marlowe's 'geographical excursions were grounded upon the latest maps of that elegant cartographer, Abraham Ortelius. His long-drawn-out digression on military tactics, in the third act of the continuation, was so well informed and up-to-date that much of it seems to have been derived from the manuscript of a book which had not yet been published, Paul Ive's *Practise of Fortification*' (36).

[5] Natolia in Tamburlaine's time was roughly equivalent to what we might call Asia Minor. According to Ethel Seaton, 'it is the whole promontory of Asia Minor, with a boundary running approximately from the modern Bay of Iskenderun eastward toward Aleppo, and then north to Batum on the Black Sea' (20).

[6] I do not substitute the term 'urinalysis' for 'uroscopy' given that the former term did not enter the English language until 1892, according to the *OED*. Urinalysis also denotes a more detached, chemical analysis of urine not practiced in the Early Modern period.

[7] Carole Rawcliffe argues that while some mistrusted uroscopic diagnosis, 'most people retained a trusting belief in uroscopy' (47). There were, of course, detractors, but these were an unheard handful. The most interesting example is the 1637 text *Pisse-Prophet* by Thomas Brian, though notably after Marlowe's time. The author objects to the science's fundamental reliance on the construction of nar-

ratives based on the both prescribed tradition and the rhetorical sophistication of drawing out the patient's condition from the urine's carrier. Thomas Brian wrote in 1637 that '(*Urina est meretrix, vel mendax*) the Vrine is an Harlot, or a Lier; and that there is no certaine knowledge of any Disease to be gathered from the Vrine alone, nor any safe judgement to be exhibited by the same' (A2^r). Brian does assert, however, that uroscopy in conjunction with other methods is acceptable. It is worth adding that few authors of surviving uroscopic manuals proclaim that uroscopy should be used to the exclusion of other diagnostic methods. Generally, as Gail Kern Paster writes, 'the physicians who read urines did not see themselves as fantasists. They were eager to distinguish their own practice in assessing the true signifying potential in urine from the grander claims of uromancers – those who read samples of urine to tell fortunes – or even from the practice of those empirics, women healers, and other lay practitioners who judged urines apart from other bodily signs and often diagnosed disease without examining the patient. Physicians inspecting urines believed themselves engaged in the scientific recovery and classification of bodily signs that were objectively *there*' (41–2).

[8] Peter Lowe in *The Whole Course of Chirurgerie* (1597) explains the spirits as follows: 'CO. What are spirits. LO. They are a substance subtill and aerious of our bodie, bredde of the part most pure and thinne of the bloud, sent through all the bodie, to the effect, the members may doe their proper actions. CO. In what partes of our bodie abound they most. LO. As sayth Andreas Lorraine, they abound in the heart and arters, in the braines and nerues. CO. How many sorts of spirits are there. LO. Three, animall, vitall & naturall. CO. What is the spirit animall. LO. It is that which remaineth in the braines, of which a great parte is sent to the eyes, by the nerues optickes, some to the eares and diuers other partes, but most to the eyes, therefore, those who haue lost their sight, haue the other vertues more strong, the which caused Democritus & other Philosophers to [p]ut out their eyes, to the end their vnderstanding might be more cleare. CO. Is the spirit animal brought through all the nerues substantially. LO. No, but onely by the nerues optickes, because they haue manifest hollownesse, and not the rest. CO. What is the spirit vitall. LO. It is that which is in the heart and arters, and is made of the euaporation of the bloud and of the ayre laboured in the lights by the force of vitall heat, and thereafter is diffused through the members for the conservation of the naturall heate. CO. What is the spirite naturall. LO. It is that which is ingendred in the liuer, and veines, and there remaineth, while the liuer maketh the bloud, and other operations naturall, the use of it is to helpe the concotion' (D3^v–D3^r).

[9] Whence the accidental heat? Johnstone Parr in his book *Tamburlaine's Malady*, suggests this is a result of his natural disposition, as influenced by astrology.

[10] According to uroscopist Robert Recorde: 'Criticall dayes be such dayes, on which there is (or may be) perceived some certain token and great alteration in the sick body, either to health or death, or continuance of sickness. What these be, more at large I will hereafter (God willing) declare in a Book peculiarly, because it requireth more largeness of words, then is meet for this place. But one thing I must

tell you, that the same dayes also be called Judiciall, but not Indiciall for the Indiciall daies are of another kind' (105).

[11] The *OED* confirms early usage of obscure to mean dark or black.

[12] What causes the rise in choler is a debate traced out by Johnstone Parr (3–23).

[13] According to Allen Debus, who provides one of the more comprehensive accounts to date of the emerging Paracelsian school, Early Modern English doctors following Paracelsian methods still employed urine as a reliable diagnostic tool (156–7).

[14] For example, John Fletcher's 1598 treatise *The Differences, Cavses, and Ivdgements of Vrine* is acknowledged in 'An admonition to the Reader' to be a compilation of earlier uroscopic treatises, Recorde's text especially included.

[15] It may be no coincidence that Tamburlaine also imagines the world divided into three: 'I will confute those blind geographers / That make a triple region in the world, / Excluding regions that I intend to trace' (I, 4.4.77–9). Tamburlaine's triple region derives from the period's T-O maps (see Sanford's thorough study) that trisected a circular world with a T-shaped cross. It could be argued that this tripartite geographic partitioning consciously or unconsciously imposed itself on uroscopic partitioning.

[16] My reading of the physician's diagnosis is sharply opposed to that of Marlovian critic Emily Bartels who claims the diagnosis is 'otherworldly' (80). The diagnosis, I argue, is very much of this world, or more accurately, the world of late sixteenth-century England.

[17] I am grateful to physician Trey Dubose, Chief Resident at the University of Florida's Shands Hospital, who provided an indispensable contemporary perspective on urology.

[18] J.S. Cunningham writes that 'the imaginative stimulus afforded by Ortelius is only matched in the play by Ovid's *Metamorphoses*' (20).

[19] Dollimore specifically means transgression in the Foucauldian sense that 'transgression contains nothing negative, but affirms limited being' (Foucault, 'Preface to Transgression' 35). Dollimore, however, does not seem to explore fully Foucault's paradoxical definition of the term, so vividly described in this passage: 'Transgression, then, is not related to the limit as black to white, the prohibited to the lawful, the outside to the inside, or as the open area of a building to its enclosed spaces. Rather, their relationship takes the form of a spiral which no simple infraction can exhaust. Perhaps it is like a flash of lightening in the night which, from the beginning of time, gives a dense and black intensity to the night it denies, which lights up the night from the inside, from top to bottom, and yet owes to the dark the stark clarity of its manifestation, its harrowing and poised singularity; the flash loses itself in this space it marks with its sovereignty and becomes silent now that it has given a name to obscurity' (35).

[20] Burnett's footnote here merely redirects us to Bakhtin and not to scenes from popular English literature of the time.

[21] Levin's classic study of the play concurs on the play's tone: 'it is not bathos but blasphemy' (39).

[22] Michael Schoenfeldt in *Bodies and Selves in Early Modern England* critiques the larger Bakhtinian critical enterprise and its domination of much theoretical exploration of the body.

[23] Kristeva writes: 'The corpse (or cadaver: *cadere*, to fall), that which has irremediably come a cropper, is cesspool, and death: it upsets even more violently the one who confronts it as fragile and fallacious chance' (3).

[24] Tamburlaine's end mirrors that of Ortelius. R.A. Skelton's introduction to the facsimile of the London edition of *Theatrum* tells us that: 'From the first edition of 1570 to that of 1595, the last to be published in Ortelius' lifetime, the contents of the *Theatrum* itself increased from 53-115 map sheets… at his death Ortelius left unpublished plates, to be incorporated in later editions by his successors' (xi). Ortelius, too, left much unconquered.

Chapter 12

'The Wronged Breeches': Cavalier Scatology

Peter J. Smith

I

Published in his *Poems* of 1680, John Wilmot, Earl of Rochester's 'Song' typifies the quintessence of the cavalier:

> I Rise at Eleven, I Dine about Two,
> I get drunk before Seven, and the next thing I do;
> I send for my *Whore*, when for fear of a *Clap*,
> I Spend in her hand, and I Spew in her *Lap*:
> There we quarrel, and scold, till I fall fast asleep,
> When the *Bitch*, growing bold, to my Pocket does creep;
> Then slyly she leaves me, and to revenge th'affront,
> At once she bereaves me of *Money*, and *Cunt*.
> If by chance then I wake, hot-headed and drunk,
> What a coyle do I make for the loss of my *Punck*?
> I storm, and I roar, and I fall in a rage,
> And missing my *Whore*, I bugger my *Page*:
> Then crop-sick, all *Morning*, I rail at my *Men*,
> And in Bed I lye Yawning, till Eleven again.[1]

This diurnal round of eating, drinking, molestation, quarrelling, theft, and exploitation seems, at first sight, to be brazenly and aggressively self-determined. The day starts when the narrator wakes and ends when he sleeps. His selfishness is inscribed on every clause; every line contains at least one use of the first-person pronoun and in no fewer than nine lines it appears twice or more. The narrator of the poem (could that be Rochester?) is unbound by any social protocols, sexual proprieties or cultural niceties.[2] Only his fear of disease influences his behavior, ejaculating in his mistress's hand instead of her vagina.[3] He sleeps in (surely a rebuttal of the Puritan work ethic), dines late and gets drunk early. He spends the evenings with his whore and, if she is unavailable, he will turn to the nearest alternative and bugger the page. The perfunctory nature of this alternative is symptomatic of the speaker's egocentricity – others exist only to satisfy his pleasure whether they be his servants against whom he rails, his whore with whom he quarrels, or the pageboy. But the boy is only a quiet, passive receptor of the speaker's sexual appetites and, we might add, fluids. Obviously the page's consent is unsolicited and unrequired as he exists merely to satisfy and, in so doing, pacify his master, called upon when all other options have proved unavailable.[4] Of all the secondary figures in the poem, it is the page who is treated most dismissively, most anonymously and most cavalierly. His being compulsorily sodomized is the final demonstration of the fiercely abusive character of the cavalier – a term which denotes a level of courtly sophistication but which has come to connote, typified by conduct like the above, haughtiness, nonchalance, arrogance, disdain, even superciliousness.

Yet the description of the daily routine of this spoiled and soiled patrician rapidly takes on an air of hollowness. The poem's largely paratactic grammar lists his reiterated daily activities in a banal inventory which intimates ennui rather than gratification. In spite of his Epicurean pursuit of the good life, what he describes is a world which turns around the sating of appetites – appetites which will require just as much assuaging the very next day. The poem makes its end where it began – 'I Rise at Eleven … till Eleven again' – and this orbital continuity suggests that this dissolution constitutes nothing more than a vicious circle of addiction inside of which the cavalier eats, drinks, spews and fucks his way to oblivion. There is nothing sensual or erotic about the cavalier's sexual exploits, nothing savoury or satisfying about the food or wine. This is a life of conspicuous but nugatory consumption which eludes satisfaction and stumbles drunkenly towards despondency and melancholy. Along with the vanity of the cavalier's lifestyle, his very identity is swamped with craving. The working out of his desires is little more than a mechanical process as the egocentricity noted above is transmuted into bleak solipsism in the alembic of despair.

This, then, is the inverse of the cavalier as royal courtier, as heroic, curly-locked swaggerer, as gifted poet and diplomat, as sensual lover. This is the flip side of Caroline ostentation, the equestrian portraits of King Charles or the relaxed splendor of Sir Anthony van Dyck's dynastic groupings. Miles away from the luxuriant pastorals of Sir Peter Lely, or the voluptuous eroticism of poets such as Sir John Suckling or Thomas Carew, this is a world, in Vivian de Sola Pinto's phrase, 'of ugliness, cruelty, and filth' (4: 147). This tripartite Hobbesian formulation is no accident. It is among others, argues Pinto, the influence of Hobbes with his 'mechanico-materialistic universe' which brings about the ontological crisis giving rise to physical impotence, theological confusion, intellectual despair and religious doubt (149). As Rochester's bedside confessor and first biographer, Gilbert Burnet, records, Rochester formulated this nihilism with disarming simplicity: 'He said, They were happy that believed: for it was not in every man's power' (quoted by Farley-Hills, *Rochester* 65).

In this dystopia, caustic dissatisfaction turns readily into self-loathing, shame, and disgust. Rochester's various poems about masculine impotence rail at his human frailty and mire his suave, cavalier aesthetic in obscenity. Admonishing his limp manhood in 'One Writeing Against his Prick,' he challenges its obdurate failure via a series of savage rhetorical questions:

> Did shee not raise thy drooping head on high
> As it lay Nodding on her wanton thigh?
> Did shee not clap her leggs aboute my back
> Her *Porthole* open? Damn'd *Prick* what is't you lack? (11–14)

and the poem ends with a rigid warning: 'Hence forth stand stiff and gaine your creditt los't / Or I'le nere draw thee, but against A *Post*' (15–16). Again, in 'Song,' the frustration of sexual impotence is powerfully conveyed by the blunt and trashy pornographic descriptions of his mistress: 'Her Leggs stretcht wide, her Cunt to me did show' (6). Yet in spite of the unabashed nature of this peep show, he is unable to attain an erection: 'Yea, thô she took it in her warm moyst hand / And Crammed it in (Dul Dog) it would not stand' (17–18). This is the cavalier as hollow man, pox-ridden, drunk, violent, and impotent. For David Farley-Hills, Rochester's verdict on love is as 'a degraded experience and, therefore, an archetypal situation for demonstrating the fatuousness, the essential meaninglessness, of all human experience' (105).

While Farley-Hills captures the poignant incoherence of Rochester's vision, his use of 'archetypal' and the generalisation of 'all human experience' erases the specific historicity of the poetry. Rochester's is not a universal disgust but, as the many references to his contemporaries suggest, it results from a precise set of historical and political circumstances. It is in his biography of Rochester that Graham Greene ascribes this revulsion to the disappointments of the Restoration:

> The sombre desperate mood in which his poetry seems usually to have been composed was aided by the general disillusionment of the time. So much had been expected of the Restoration and so little had come of it; so much had been expected of the war with Holland, and it brought only disgrace. Vast hopes disappointed are apt to leave an overmastering weariness behind, and war is not usually followed by peace of mind. The excitement of war is a drug that we miss in the dull days after. Spleen was the disease which ran through the literature of the succeeding quieter years.... Disillusionment, monotony, boredom preyed worst on the finest spirits. (74)

Remarkable is the way in which Greene describes political turmoil in relation to physical illness. The entire culture of the day, according to Greene, was splenetic. Gordon Williams echoes this and, in the introduction to his magisterial *Dictionary of Sexual Language and Imagery in Shakespearean and Stuart Literature*, he attributes Restoration disgust to the humiliation of Republican victory in the English Civil War. Rochester's poetry, composed around the time of the Restoration, is physically scarred by the experience of defeat; Williams writes: 'What the sexual language of Restoration court society illustrates above all is a maimed elite, still reeling from the shock of being challenged – and defeated – by a bunch of farmers and shopkeepers' (1: viii–ix). More pertinent still, for our purposes, is Williams's description of the nauseating aesthetic that follows the overthrow of the royalist body politic: 'there is a new dimension to the harping on natural emissions, which become the repulsive suppurations of disease' (1:ix).

Fittingly, this 'harping on natural emissions' is evident in Burnet's descriptions of Rochester's own decrepitude:

> He had not been long in the Country when he thought he was so well, that being to go to his Estate in *Somersetshire* he rode thither Post. This heat and violent motion did so inflame an Ulcer, that was in his Bladder, that it raised a very great pain in those parts: Yet he with much difficulty came back by Coach to the Lodge at *Woodstock-Park*. He was then wounded both in Body and Mind: He understood Physick and his own Constitution and Distemper so well, that he concluded he could hardly recover: For the Ulcer broke and vast quantities of purulent matter passed with his Urine. But now the hand of God touched him, and as he told me, It was not only a general dark Melancholy over his Mind, such as he had formerly felt; but a most penetrating cutting Sorrow. So that though in his Body he suffered extream pain for some weeks, Yet the Agonies of his Mind sometimes swallowed up the sense of what he felt in his Body. (quoted in Farley-Hills, *Rochester* 79)

Here Burnet insists on the contiguity of physical and mental maladies; bodily disease is inseparable from spiritual and emotional dis-ease. As Burnet later reports, 'In this disposition of Mind did he continue all the while I was with him, four days together; He was then brought so low that all hope of Recovery was gone. Much purulent matter came from him with his Urine, which he passed always with some pain; But one day with unexpressible torment' (quoted in Farley-Hills, *Rochester* 85–6). The force of the metaphor of the body politic is thus reified in Rochester's own corporeal decline. It is as though he figures forth in the concoction of puss and piss, the crisis of the aristocracy displaced from the centers of power by the middling sort.

As Jeffery Persels has noted in respect of an earlier period, 'the proper functioning of the body human has direct bearing on the proper functioning of the body politic .... The body, the intestines, excrement and their relative health thus prove to be so many evident metaphors for social order' ('Straitened' 105). This, of course, is only half the story as the health of the nation is just as likely to be the vehicle as the tenor of the metaphor. Physical affliction need not just be symbolic of social disharmony, but it can *be symbolized by* social disharmony. Rochester's depiction of the body politic being subjected to the whims and incontinences of the body natural (as in the much vaunted debauchery of Charles II himself) becomes then a discourse of political critique (at least potentially) and issues from within its own ranks. Proximity to the center makes the realization that it cannot hold all the more disturbing. As Pinto puts it, 'The criticism is all the more effective because it comes from one who is within the society which is criticized and who possesses its virtues of easy sureness of tone and diction and poise of gesture and rhythm' (146). Such a regime, Rochester insists self-consciously, has lost its potency and cavalier corporeality becomes infused with the despondency of failure, riddled with the disfiguring diseases of political disgrace.

Unsurprisingly, the kind of scatology we witness in Rochester shares this darkness of tone and is often indistinguishable from the sexual disgust we have already seen. Rochester stages a poetic exchange between Louise de Keroualle,

Duchess of Portsmouth, and Nell Gwyn (both mistresses of Charles II) in which they rail at each other and agree only in their joint damnation of Hortense Mancini, duchesse de Mazarin, who, following her arrival in December 1675, was also vying for the king's attentions:

NELL: When to the King I bid good Morrow,
With Tongue in Mouth, and Hand on Tarse,
Portsmouth may rend her Cunt for Sorrow,
And *Mazarine* may kisse myne *Arse*.

PORTS: When Englands *Monarch*'s on my Belly
With Prick in Cunt, thô double Cramm'd,
Fart of mine Arse, for small whore Nelly
And Great Whore *Mazarine* be damn'd. (1–8)

This obscene dialogue exemplifies the mutual assimilation of sexual and scatological disgust. The poem vacillates between the brusque anality of Nell's and Portsmouth's curses ('kisse myne *Arse*' and 'Fart of mine Arse') and the casually gross descriptions of sexual activity. Yet the poem is more than mere bathos. It would seem to demonstrate the historically specific abjection and despondency which Greene, Williams, and Pinto describe and which we have seen writ large in 'Song' ('I Rise at Eleven…'). Nell, Portsmouth and the absent Mazarin are the unfortunate heirs of a tradition of literary misogyny seen at its most vicious in the tragic dramas of the Jacobeans and, like the women in Tourneur or Webster, they are synechdochic figures of a historically localized cultural decline. Their filthy language, straddling the sexual and the scatological, fuses one orifice with another: cunt and arse become synonymous as their corruption is figured both front and back. Elsewhere, the speaker of 'The Advice' encourages the shepherd to abjure the company of women on the grounds that they are more trouble than they are worth. The female capacity to provide release from sexual tension is nothing more than 'Easing us of an Excrementish Load' (15). While the term *excrement* could refer to different kinds of innocuous physical product such as hair, nails, or even feathers, the word was also being used to refer specifically (according to *OED* at least since 1533) to fecal matter.[5] Rochester's use of 'Excrementish Load' draws together vaginal and anal as the ultimate destination of the shepherd's ejaculate.

But perhaps the most explicit superimposition of front and back orifices and one which is most historically contingent on the recent civil strife occurs in a lyric of eight stanzas which recounts the double penetration of Phyllis 'In the *Fields* of *Lincolns Inn*.' While Coridon's penis is 'fitted, / To her less frequented Arse' (11–12), Strephon's is thrust 'into her *Cunt*' (16). What follows is an extraordinary account of a penile battle which, within the body of Phyllis, rehearses the imbroglio of political disorder:

Now for Civil *Wars* prepare,
Rais'd by fierce intestine bustle,

When these *Heroes* meeting justle,
In the *Bowels* of the fair.

They tilt, and thrust with horrid pudder,
Blood, and slaughter is decreed;
Hurling Souls at one another,
Wrapt in flakey *Clotts* of *Seed*. (17–24)

This internal dissension, strife within the '*Bowels* of the fair,' figures forth the historic as hysteric (in its Greek sense of 'belonging to the womb'). The 'fierce intestine bustle' literalizes the metaphor of the body politic so that Phyllis's vagina / anus becomes the battle ground for the warring weapons of her suitors. But note also the martial aspect of the struggle, the heroes jostle one another, tilting, thrusting 'with horrid pudder,' and the inevitable products of battle, blood, and slaughter, are fatally determined while the spilling of souls, manifest in their emissions, both vaginally and anally, is simultaneously an image of death and the little death of orgasm.[6] The role of Phyllis in the performance is worth examining. Though she initiates the activity – the second stanza informs us that it is she who 'Lay most pensively contriving, / How to Fuck with *Pricks* by pairs' (7–8) – once the shepherds take up battle stations, she is pretty much absent, merely the field on which the combat is being fought. Her absence turns the business into something very like a homoerotic exchange or at least a homosocial encounter, a tournament between knights for the honour of an observing but passive lady rather than a sexual feat in which she takes an active role. In this way the poem again draws together cunt and arse; if the battlefield is a level one (though Strephon's prick 'was a handful longer' [13]), Phyllis becomes merely the provider of two apertures rather than an erotic or aroused presence in her own right. Oddly, at this point, the poem insists on a kind of internal propriety which prevents the mutual annihilation of the two warriors. Nature's barrier, the wall of the womb, ensures that martial strife remains only metaphorical. For all the vigor of their warfare, both pricks live to fight another day:

Nature had 'twixt *Cunt* and *Arse*,
Wisely plac'd firm separation;
God knows else what desolation
Had ensu'd from *Warring Tarse*. (25–8)

Nature's mission is to protect the warring tarses (or as Rochester has the plural, 'Tarse') no matter which hole they have occupied.[7] But the uterine wall is not intended to forbid or even discourage anal intercourse; rather its role is to pacify the warring cocks and guarantee their future potency. Even as the poem insists on the separation of front and back, then, this is not the primary objective. In fact, the conclusion of the poem delineates a mutual satisfaction without identifying a victor. All are sated no matter where they have been: 'The *Nymph* was sorely *Ballock* beaten, / Both the *Shepherds* soundly tir'd' (31–2). The fact that the shepherds are undifferentiated implies the interchangeablility of their coital destinations –

they may be divided from each other, front and back are equivalent objectives.

Another example of this unification of sexual and fecal occurs in Rochester's description of 'Mrs Willis.' The fifth stanza runs together the one and the other in language that is disarmingly platitudinous, almost axiomatic:

> Bawdy in thoughts, precise in Words,
> Ill natur'd though a Whore,
> Her Belly is a Bagg of Turds,
> And her Cunt a Common shore. (17–20)

The body's interior, that which should be numinous as the location of the soul is here graphically described as the seat of the body's corruption. Drawing on the medieval *contemptus mundi* motif, the body is seen to be nothing more than a site of constant decay – 'Remember oh Man that dust thou art and unto dust shalt thou return.' In *Christes Tears over Jerusalem*, Thomas Nashe provides an especially malignant example of this trope when he describes prostitutes as 'excrementall vessels of lust' and his aggressive questioning metamorphoses the whores into lavatories: 'What are you but sincks and privies to swallow in mens filth?' (quoted in Paster 154). Unsurprisingly the same trope is found in Rochester; in 'Broad Side against Marriage' he refers to a whore as the 'Close stool to Venus' (76), running together ideas of arousal and defecation as though both are in similar need of release. This homology, according to which front and back orifice are rendered synonymous, is the final demonstration of the inherent corruption of mankind born *inter urinas et fæces*. As Nicholas Culpeper notes, 'The neck of the womb is seated between the passage of Urin and the right Gut, to shew fond man what little reason he hath to be proud and domineer, being conceived between the places ordained to cast out excrements, the very sinks of the Body' (quoted in Paster 211). This vehicle for religious self-abasement – the motif can be traced back to the writings of the biblical exegetes – is secularized by Rochester. The poet is clearly less interested in ethical meditation or spiritual crusading than he is in damning his political enemies and using them to represent more specifically historical flaws. What was a convention of medieval religiosity is here transformed into a malevolent barb; Rochester's assimilation and transformation of this scatological device is, in every sense of the word, cavalier.

The contrast between this Restoration disgust and earlier carnivalesque scatology can be seen in the distance – in substance and levity – between Rochester's employment of the device, and the birth of Gargantua. In Sir Thomas Urquhart's translation of Rabelais's *Gargantua and Pantagruel*, Gargamelle's anal prolapse appears as follows:

> … a little while after she began to groane, lament and cry, then suddenly came the midwives from all quarters, who groping her below, found some *peloderies*, which was a certaine filthy stuffe, and of a taste truly bad enough; this they thought had been the childe, but it was her fundament, that was slipt out with the mollification of her *streight intrall*, which you call the *bum-gut*, and that meerly by eating of too many tripes, as we have shewed you before: whereupon an ugly

> old trot in the company, who had the repute of an expert she-Physician, and was come from *Brispaille* near to Saint *Gnou* three-score years before, made her so horrible a restrictive and binding medicine, and whereby all her *Larris*, arse-pipes and conduits were so opilated, stopped, obstructed, and contracted, that you could hardily have opened and enlarged them with your teeth, which is a terrible thing to think upon ... by this inconvenient the *cotyledons* of her matrix were presently loosed, through which the childe sprung up and leapt, and so entering into the hollow veine, did climbe by the diaphragm even above her shoulders, where that veine divides it self into two, and from thence taking his way towards the left side, issued forth at her left eare. (37)

Paradoxically, while Rabelais relishes the details of the prolapse, and the foul-smelling and tasting excrescences which result therefrom, the baby is not borne *inter urinas et fœces*. Instead, in a parody of the birth of Athene from the head of Zeus, he issues from Gargamelle's left ear, untainted by the stench and soil pouring from his mother's lower bodily strata. This is a miraculous birth: 'as soone as he was borne, he cried not as other babies use to do, *miez, miez, miez, miez,* but with a high, sturdy, and big voice shouted aloud, Some drink, some drink, some drink, as inviting all the world to drink with him' (37–8). Gargantua's instant invitation to 'all the world' contrasts with the private consumption of the cavalier courtier. On the one hand, Rochester consumes food, wine and boy in the privacy of his own room while he voids himself secretly in the appropriately named 'privie.' In Rabelais, on the other hand, as Mikhail Bakhtin notes, the banquet images constitute a 'popular-festive form.... This is no commonplace, privately consumed food and drink, partaken by individuals. This is a popular feast, a 'banquet for all the world'' (*Rabelais* 278). Even Gargamelle's evacuation is a riotous public spectacle. As Bakhtin insists, carnival is collective, commensal, and all-embracing: 'Carnival laughter is the laughter of all the people' (11). Moreover, it is the reader of the novel or the spectator of the drama who is embraced and folded in by the world of carnival: 'Carnival does not know footlights, in the sense that it does not acknowledge any distinction between actors and spectators.... Carnival is not a spectacle seen by the people; they live in it, and everyone participates because its very idea embraces all the people' (7). Rabelaisian scatology, as Bakhtin is at pains to point out, is a version of fecundity because excrement 'is linked to the generating force and to fertility ' (175). The birth of Gargantua illustrates a 'material bodily affluence, a generating and growing superabundance' (221). Anus and vagina are linked in their productivity as two places from whence the body can demonstrate its productiveness. Bakhtin proposes that, in the Early Modern period, 'excrement was conceived as an essential element in the life of the body and of the earth in the struggle against death. It was part of man's vivid awareness of his materiality, of his bodily nature, closely related to the life of the earth' (224). Little wonder, then, that the excrement Gargamelle expels is so readily taken for the new-born babe.

Cavalier poetry, by contrast, runs together back and front, male and female in an orgy of confusion and degradation:

Since Fucking is not, as 'twas wont,
The Ladyes have got a new Trick,
As an Arsehole serves for a Cunt
Soe a Clitoris serves for a Prick. (9–12)

This sexual and fecal imbroglio, as we have already suggested, is a sign of the times, these lines appearing in a poem which opens, 'Oh! what a damn'd Age do we live in.' Writing less than eighty years after Rochester's death, the Scottish philosopher, David Hume, attributes the poet's obscenity to the depravity of his age: 'The very name of Rochester is offensive to modern ears; yet does his poetry discover such energy of style and such poignancy of satyre, as give ground to imagine what so fine a genius, had he fallen *in a more happy age* and followed better models was capable of producing' (quoted by Farley-Hills, *Rochester* 200, emphasis added). The jubilant kind of anality and sexual fecundity which we witness in Rabelais is extirpated as Rochester's focalized hostility directs it to a local satirical target – his own peers, his own society, and even particular named individuals as we have seen. Cavalier scatology is infused with a political specificity which is less pronounced in that of the earlier period.

Commonly in Rochester, the anus and the vagina are blurred together in the act of sodomy. In 'Song' (beginning 'I Rise at Eleven…'), as we have already seen, the poet will spend in his page's anus if his whore's vagina is unavailable. 'The Disabled Debauchee' is almost indifferent to the identity of his sexual partner as he, his mistress Cloris, and the linkboy (that is, a torch-bearer) satisfy themselves on each other indiscriminately:

Nor shall our Love-fits *Cloris* be forgot,
When each the well-look'd Linkboy strove t'enjoy;
And the best Kiss was the deciding Lot,
Whether the Boy Fuck'd you, or I the Boy. (37–40)

The rake's disinterested attitude makes whore and boy equal as sites of (vaginal or anal) consumption.

Elsewhere the fusion of orifices extends to the blending of their respective products. Discharge from either opening might make intercourse ill-advised:

By all *Loves* soft, yet mighty *Pow'rs*
It is a thing unfit,
That *Men* shou'd Fuck in time of *Flow'rs*
Or when the *Smock*'s beshit. (1–4)

The 'nasty *Nymph*' is implored to use 'Paper still behind, / And Spunges for before' (7–8). But in a much more disturbing poem, published for the first time in 1999, we hear of a mysterious witch:

… deep red in Charms, and Spells;
Philters, and *Potions*, that by *Magick* skill,

> Can give an *Eunuch* Stones, and *Cunt* its fill.
> *Babes*, at her call fly from the breeding *Womb*,
> With *Neighbor Turd*, in loathsome *Jakes* to roame. (21–5)

The hag, a backstreet abortionist, deposits '*Babes*' and '*Neighbor Turd*' in the '*Jakes*' (or toilet) so that the kind of anal Rabelaisian fertility we considered above is here horribly inverted. The poem goes on to detail how unforthcoming women will be revenged by having one bodily orifice transformed into the other in a metamorphosis of staggering obscenity: '*For this abuse, the* Rump-fed-Runts *shall mourn, / Till slimey Cunt, to grimey Arse hole turn*' (71–2). What we see in Rochester's cavalier scatology is the deliberate obfuscation of sexual categories and a refusal to preserve the distinctions between comic scatology and bleak, predatory sexuality.

II

It is nearly twenty years since Francis Barker, catching Pepys, in the privacy of his own chamber (like the cavalier's private place) masturbating over a lewd book, developed his ground-breaking formulation about the emergence of bourgeois subjectivity. This act of repression, what Barker calls the 'enclosure of the Pepysian moment,' forces subjectivity in upon itself and superimposes autonomy and solitude:

> The diagram of the text [that is, the *Diary*] is as a series of concentric circles at the furtive heart of which is the secret declivity of the soul itself. The I surrounded first by discourse, then by the *domus*, the chamber, and finally by the public world, is placed at the heart of its own empire, in silence and very largely in terror. The *Diary* for all the fullness of its days, despite being so richly populated with others and with the furniture of gossip and events, is thus the record of a terrible isolation. (9–10)

This isolation, which will produce the self-loathing of Rochester's cavalier, is as we have already seen, and as Barker asserts once more, the outcome of recent historical upheaval: 'individuation is historically produced – this situation is the result of the revolutionary process that preceded it. The political upheaval of the mid-century established, as all revolutions must if they are to be thoroughgoing, a new set of connections between subject and discourse, subject and polity' (10). For our purposes, the most significant aspect of this epistemological shift was the establishment of 'new images of the body and its passions [which] were a crucial, if increasingly occluded, element' (10). How fitting that such new cultural repression of corporeal operation should find expression through the form of the diary.[8] As Norman O. Brown puts it, 'only repressed life is in time, an unrepressed life would be timeless or in eternity' (93). The diary is the most rigorously ordered genre chronologically and, in Pepys, as Barker demonstrates, it charts the very moment 'when the division between the public and the private is constructed in its modern form' (14).

It is in terms of the treatment of his own body that Pepys voices this historically novel attitude to the body – an attitude which, not surprisingly, sounds very much like anxiety. On the morning of 7 October 1663 Pepys woke with a 'pain that continued on me mightily, that I keeped within all day in great pain, and could break no wind nor have any stool after my physic had done working. [Later that day] having a good fire in my chamber, I begun to break six or seven small and great farts; and so to bed' (4: 327). This contentment is short-lived though. We hear only two days later that 'I could neither have a natural stool nor break wind, and by that means still in pain and frequent offering to make water. [I] do not think myself likely to be well till I have a freedom of stool and wind' (9 October 1663; 4: 328).[9] The pains are not only anal but genital (in an uncanny adumbration of Rochester's amalgamation of back and front, the other side of the *jouissance* that is Freud's 'polymorphous perversity'): 'the greatest of my pains I find to come by my straining to get something out backwards, which strains my yard and my cods, so as to put me to a great and long pain after it .... After supper to bed as I use to be, in pain, without breaking wind and shitting' (11 October 1663; 4: 329–30).[10] Pepys resolves to tackle this constipation: 'I see I must take, besides keeping myself warm, to make myself break wind and go freely to stool before I can be well' (4: 332). To this end, he calls upon his wife to administer an enema:

> Anon about 8 a-clock, my wife did give me a Clyster which Mr. Hollyard directed, *viz.*, A pinte of strong ale, four ounces of Sugar, and two ounces of butter. It lay while I lay upon the bed above an hour, if not two. And then, thinking it quite lost, I rose; and by and by it begun with my walking to work, and gave me three or four most excellent stools and carried away wind – put me into excellent ease; and taking my usual wallnutt quantity of Electuary at my going into bed, I had about two stools in the night – and pissed well. Voided some wind. (4: 332)

The delayed effects of the enema offer an instance of the puzzlement caused by the bowel. While digestion is extensively theorized as the center of bodily health, Pepys assumes the worst, that nothing will come of it; he thinks 'it quite lost.' The eventual farts and movements come as a pleasant surprise (hence his use of 'excellent' to describe his stools).[11] If on the one hand the body is beginning its trajectory towards individuality and even privacy (at its extreme a Rochesteresque solipsism), it is still sufficiently 'other' to elude the control of its narrator.

This inability to regulate the body makes Pepys reliant on his doctor and his wife and even at the mercy of divine intervention: 'I have by my wife's good advice almost brought myself, by going often and leisurely to the stool, that I am come almost to have my natural course of stool as well as ever, which I pray God continue to me' (11 December 1663; 4: 414). Pepys is also happy to consult his workmates and evidently felt no embarrassment seeking the advice of high-ranking civil servants and Admiralty big-wigs:

> I did however make shift to go to the office, where we sat; and there Sir J. Mennes

> and Sir W. Batten did advise me to take some Juniper water, and Sir W. Batten sent to his Lady for some for me, strong water made of Juniper. Whether that, or anything else of my draught this morning did it, I cannot tell, but I had a couple of stools forced after it and did break a fart or two. (10 October 1663; 4: 329)[12]

The first of the colleagues mentioned here, Sir John Mennes (to whose own poetry we shall shortly turn) seems to have taken a keen interest in matters scatological. On 25 September 1662 Pepys records how Mennes 'told us, among many other things, how in portugall they scorn to make a seat for a house of office. But they do shit all in pots and so empty them in the river' (25 September 1662; 3: 205). But it was to Pepys especially that Mennes offered his advice about bowel movements, sometimes, if we are to believe Pepys's irritated tone, with officiousness: 'Up and to my office, where all the morning – and part of it Sir J. Mennes spent as he doth everything else, like a fool, reading the Anatomy of the body to me' (10 October 1663; 4: 334). Mennes, born in Kent at the end of the 1590s and thirty-five years Pepys's senior, typified the culture from which the new subjectivity was in the process of emerging; in Barker's words, he hailed from the 'older sovereignty of the Elizabethan period [which, following the upheaval of the Civil War, had been] disassembled' (10).[13] Mennes thus provides an ideal test case against which to try the hypothesis that changing attitudes to scatology are symptomatic of a shift in sensibility brought about by the crisis of the embattled aristocracy. Thus, if Pinto, Greene, Williams, and Barker are correct, albeit in their various ways, in asserting that post-Restoration obscenity is torridly intensified by the overthrow of the Cavalier cause, then we might expect to find a less pessimistic scatology in the years leading up to the Republican triumph.

Turning to *Musarum Deliciae or The Muses Recreation* which appeared a quarter of a century before Rochester's *Poems*, such a hypothesis does indeed seem to be borne out. *Musarum Deliciae*, published in London in 1655, is, nowadays, a little-known collection. Its compilers, Mennes (Pepys's unofficial proctologist) and James Smith are literary nonentities save for the excellent study on *Cavaliers, Clubs and Literary Culture* by Timothy Raylor in which he takes this text as exemplifying the processes of collaborative composition which typified the literary scene in mid-century London. Raylor suggests that modern critical indifference to Cavalier poetry is directly a result of its obscure topicality or political interestedness and, indeed, we have seen this to be the case in respect of Rochester's verse: 'the literature of the period has generally been dismissed by critics as too timebound and partisan to be worthy of serious attention' (19).

The remainder of this essay examines the kinds of scatology that appear in *Musarum Deliciae* in an attempt to demonstrate that it partakes more of the Rabelaisian spirit than the Rochesteresque; that it is a more benign, comic form of scatology without direct political targets. My title, 'The Wronged Breeches,' comes from a mock-epic poem entitled 'To a friend upon a journey to Epsam Well,' impossible to cite here in full as it runs to over 120 lines. The purgative effects of Epsom salts have been well-known for some time: 'Epsom and Glauber salts … act mainly by increasing osmotic pressure, but perhaps also by irritating the secretory cells of the intestinal lining' (Lewin 104). Also, in his *Diary* entry for 21 August 1664 Pepys records, 'So up and drunk three bottles of Epsum-water, which wrought well with me' (5: 249). In the poem two friends undertake a journey to Epsom to take the waters ostensibly for medicinal purposes:

> 'Tis here the people farre and neer
> Bring their Diseases, and go clear.
> Some drink of it, and in an houre,
> Their Stomach, Guts, and Kidneyes scower:
> Others doe Bathe, and Ulcers cure,
> Dry Itch, and Leprosie impure;
> And what in Lords you call the Gout,
> In poor the Pox, this drives all out. (23–30)

We should note in passing the poem's awareness of the manner in which different social classes are stigmatized differently. While the poor are acknowledged to have the pox, Lords have the same condition euphemized as 'the gout.'

The poem then moves on to describe the place while the miraculous powers that reside there are quickly eclipsed by the fecal depiction:

> Close by the Well, you may discerne
> Small shrubs of *Eglantin* and *Fern*,
> Which shew the businesse of the place;
> For here old *Ops* her upper face
> Is yellow, not with heat of Summer,
> But safroniz'd with mortall scumber. (31–6)

The yellow stains of the place are but a foreshadowing of the debris and detritus that will remain after the patients have left. But the allusion to 'old *Ops*' (a Roman goddess of fertility) infuses the atmosphere with a Classical seriousness which is to be debunked by the mention of *scumber* which the *OED* defines as 'Of a dog or fox: To evacuate the feces. Also *jocularly* of a person.' The coupling of Classical deity with animal discharge (intensified by the use of 'mortall') bathetically deflates the pastoral setting, traditionally taken to be a place of spiritual contemplation, with its 'Small shrubs of *Eglantin* and *Fern*.'

This comic rupturing of the Classical precedent (and, along with it, the world of Humanist emulation of the Classical ideal) continues. In particular, the most distinguished Classical sources are reduced to bumwipe (possibly an allusion to Gargantua's experiments with different abstergent objects in Chapter 13 of Book I):

> But then the pity to behold
> Those ancient Authors, which of old
> Wrote down for us, Philosophy,
> Physick, Musick, and Poetry,
> Now to no other purpose tend,
> But to defend the fingers end.

Here lies *Romes Naso* torn and rent,
New reeking from the Fundament:
Galens old rules could not suffice,
Nor yet *Hippocrates* the wise.
Not teaching, how to clense, can doe,
Themselves must come and wipe it too.
Here did lye *Virgil*, there lay *Horace*,
Which newly had wip'd his, or her Arse.
Anacreon reeled too and fro,
Vex'd, that they us'd his papers so.
And *Tully* with his Offices,
Was forc'd to doe such works as these. (37–54)

The debasement and befouling of Classical sources symbolizes the potential degradation of all that mattered to Renaissance Humanism; there is something humbling about seeing this library reduced to toilet paper. However, the tone is in stark contrast to Rochester's mordant vision. For instance, '*Tully* with his Offices' – a reference to *De Officiis* by Marcus Tullius Cicero, one of the models of Renaissance rhetoric – is clearly a scatological joke since the *OED* defines *offices* as 'the function or action of discharging excrement' (dating from 1386). There is a deliberate ambiguity being sustained between the rarefied cultural capital of the sources cited and the lowliness of their fate, piled on a dunghill, having been used merely 'to defend the fingers end.'

Even the efforts of the contemporary lover, his protestations and sonnets, end up on the midden:

Here lies the Letter of a Lover,
Which piece-meale did the thing discover
Sonnets halfe written would not stay,
But must necessity obey.
This made us for a while to think,
The Muses here did seldome drink:
But hap what would, we light from stirrup,
And streight descend to drink the syrrup. (55–62)

These Orlando-like sonnets, which had enjoyed their heyday over half a century earlier, demonstrate the presumption of the lover and confirm the sad irony of Yeats's claim that 'Love has pitched his mansion in / The place of excrement' ('Crazy Jane Talks with the Bishop' 15–16). In Rochester's 'An Epistolary Essay,' by contrast, bad poetry is not merely used to wipe the arse but becomes fecal in itself – literally fart and shit:

Perhaps ill Verses ought to be confind,
In meer good breeding, like unsav'ry wind:
Were reading forc't, I shoud be apt to think,
Men might no more write scurvily then stink;
But 'tis your choice whether you'l read or no.

If likewise of your smelling it were so:
I'd fart just as I write, for my own ease
Nor shoud you be concernd, unless you please.
Il'e own that you write better then I do,
But I have as much need to write as you:
What tho'th'Excrements of my dull brain
Runs in a harsh insipid strain,
Whilst your rich head, eases itself of wit,
Must none but Civit Catts have leave to shit? (30–43)

The contrast is fundamental: in 'Epsam *Well*' the function of poetry at the present time is 'to defend the fingers end' (42), and so artistic ambition, erudition, and Classical precedent are mocked. Intellectual achievement offers physical comfort but the lavatorial reification of artistic insight is humorous and bathetic. In Rochester, on the other hand, artistic inspiration is quite simply 'th'Excrements of my dull brain': poetic achievement is shit, pure and simple. Writing, like farting, provides alleviation – 'I'd fart just as I write, for my own ease' – while his addressee's head 'eases itself of wit.' Moreover, evacuation, is a natural necessity: 'I have as much need to write as you.' Commenting on Rochester's satires, Farley-Hills recognizes the significance of this equivalence: 'Poetry is both sublime and absurd, it is [the] psychological equivalent of ejaculation or defecation' (204).

Returning to 'Epsam *Well*,' the speaker and his companion next make their way towards the source of the spring where they are greeted and waited on by one of the ancient sentries:

The good old Father takes a cup,
When five times wash'd, he fills it up
With this priz'd Liquor, then doth tell
The strange effects of this new Well.
Quoth he, my friends, though I be plaine,
I have seen here many a goodly train
Of Lords and Ladies, richly clad,
With Aches more then ere I had:
These having drunk a week, or so,
Away with health most jocund go:
Mean while the Father thus did prate,
We still were drinking as we sat;
Till Gut by rumbling, us beseeches,
My Boyes, beware, you'l wrong your Breeches.
Ah, doth it worke? the old man cryes,
Yonder are brakes to hide your thighes.
Where, though 'twere neer we hardly came,
Ere one of us had been too blame. (63–80)

In the nick of time they feel the effects of the purgative waters and, in a parody of the ancient epic games, they take part in a shitting competition encouraged by the achievements of ancestral generations:

No wrestling here, Limbs to abuse,
But he that gaines the glory here
Must scumber furthest, shite most clear.
And, for to make us emulate,
The good old Father doth relate
The vigour of our Ancestors,
Whose shiting far exceeded ours.
...
This speech of his me netled;
With that my head I straightway put
Between my knees, and mounting scut,
At chiefest randome, forty five,
With Lyon's face, dung forth I drive,
The ayre's divided, and it flyes,
Like *Draco volans* through the skies.
Or who had seen a Conduit break,
And at the hole with fury reak:
Had he but hither took the paine
To come, had seen it once againe.
Here *Colon* play'd his part indeed,
And over-shit the stones a reed.
Whereat the Father, all amaz'd,
Limps to the place, where having gaz'd
With heav'd up hands, and fixed eyes,
Quoth he, dear, let me kisse those thighes
That prop the taile will carry hence
Our glory and magnificence. (81–112)

The poem ends in triumph and a celebration in the pub.

These extracts serve to demonstrate a definite tonal affinity with Urquhart's Rabelais and corresponding distance from Rochester in the light of which it is surely more than coincidental that Urquhart's translation of *Gargantua and Pantagruel* was published in 1653, only two years before *Musarum Deliciae*, while there remained a further twenty-five years before the publication of Rochester's *Poems*. Curiously, but just like Rabelais (or Urquhart), the knock-about humor implies a popular audience while the literary pastiche, the mock-epic and the epic catalog of asswipe authors suggest an educated or, at least, erudite one. But this tension between low and high cultural forms is also one of the distinguishing features of Urquhart's translation and one of the most significant achievements of Rabelais's *Gargantua and Pantagruel*. Raylor suggests that the catalog of Classical authors in 'Epsam *Well*' 'establishes a disjunction between the assumed value of the classics and their actual employment' although he goes on to argue that there is no threat to the status of the classics; the passage takes for granted their elevated status and is comic 'precisely because it is an aberration' (131). The epic form, epic authorities, and all that they stand for in terms of Humanist aspiration, are never undermined. Raylor concludes that 'Epsam *Well*' is 'predicated upon a commitment to concepts of ceremony, classicism, good-fellowship, and social order' (135). It would be difficult to find a more obvious contrast with Rochester in terms of tone and ethical value.

Between the farcical Interregnum scatology of *Musarum Deliciae* and the bleak Restoration scatology of Rochester's *Poems* occurred an irrevocable shift in sensibility. This perhaps is most clearly seen in the anxieties voiced by Pepys's *Diary* which contains, cheek by jowl, both a candid openness about matters scatological as well as an incipient embarrassment. *Musarum Deliciae* offers perhaps the last articulation of what we might think of as a golden age of 'innocent scatology' which looks back ultimately to Chaucer. Once that innocence is lost, and Rochester's work, as I have demonstrated, signals its final extinction, the introduction of a sharper political agenda lends the scatology a darker tone which looks forward ultimately, of course, to the writings of Jonathan Swift.

Notes

[1] I am grateful to Jeff Persels and members of the 1999 Sixteenth Century Studies Conference panel for suggestions as well as to Diana Barnes and Richard J. Larschan who read an earlier version of this paper and Michael Davies who shared his work on Rochester with me. For advice about Renaissance midriffs, my thanks go to Eleanor Lowe. William Smerdon (usually around dinner time) explained the ins and outs of proctology.

[2] Michael Davies has noted the immanence of Rochester in his own work. Citing one of the poems which names Rochester in the final line, he writes: 'No matter how self-ironizing a poem like 'To the Post Boy' might finally be, nevertheless it presents a portrait of Rochester's decadence which is simultaneously laughable in its outrageousness and horrifying in its depravity' (46). Writing of 'Ramble in St James's Park,' Farley-Hills asserts that 'poet and protagonist are clearly related' (204).

[3] On the dangers of venereal disease, see Rochester's 'Advice to a Cuntmonger' which begins: 'Fucksters you that would bee happy / Have a care of *Cunts* that Clapp yee' (269–70).

[4] The cynical use of the pageboy darkly anticipates that of the typist in Eliot's grim epic as 'the young man carbuncular … / Endeavours to engage her in caresses / Which still are unreproved, if undesired' (*The Waste Land* 231-38). Note how Eliot has parodically embedded this sordid seduction within the form of an English sonnet (235-48).

[5] See, for instance, Shakespeare's non-fecal use of the term in *Comedy of Errors* 2.2.79, and, more ambiguously, in *Love's Labours Lost* 5.1.109.

[6] Shakespeare, too, dwells with prurience on this conflation of spirit and ejaculate in Sonnet 129 'Th' expense of spirit in a waste of shame….'

[7] The *OED* defines *tarse* as 'the penis,' deriving from Old English. Gordon Williams notes that in Rochester's *Sodom*: 'The king of Gomorrah […] is named Tarsehole; and the author of *Juvenalis Redivivus* (1683) resolves to 'quit

Tarsander's Praise,' commenting: '*Those that have read* E. *of* R. *Poems, know very well what I mean by* Tarsander' (3: 1366).

[8] Incidentally, the *Diary* contains several references to Rochester, none of them complimentary. Pepys is scandalized by Rochester's abduction of Elizabeth Malet, an episode which landed the errant Earl in the Tower. See the entry for 28 May 1665 (6: 110). On 17 February 1669 Pepys records how Rochester gave T. Killigrew 'a box on the ear in the King's presence' and is offended that the king readily pardoned Rochester, noting sternly that it is 'to the King's everlasting shame to have so idle a rogue his companion' (9: 451–2). On another occasion Pepys overhears the court gossip – 'a story of my Lord of Rochester's having of his clothes stole while he was with a wench, and his gold all gone but his clothes found afterward, stuffed into a feather-bed by the wench that stole them' (2 December 1668; 9: 382). Pepys kept his own copy of Rochester's lewd poetry in a drawer in his desk. He had it bound with Burnet's *Life and Death of Rochester* in such a way that only Burnet's title is visible on the spine (10: 488).

[9] Three years earlier, Pepys records a bizarre account of how he was able to make water, all too well: 'This night I had a strange dream of bepissing myself, which I really did; and having kicked the clothes off, I got cold and found myself all muck-wet in the morning and had a great deal of pain in making water, which made me very melancholy' (28 May 1660; 1: 162).

[10] Pepys's testicles seem to have caused him some problems: 'One evil more I have, which is that upon the least squeeze almost, my cods begin to swell and come to great pain, which is very strange and troublesome to me; though upon the speedy applying of a poultice, it goes down again and in two days I am well again' (26 March 1664; 5: 98).

[11] The inability to fart is a sure prognosis of discomfort: 'a sure præcursor of pain, I find, is sudden letting off of some farts; and when that stops, then my passages stop and my pain begins' (20 June 1664; 5: 183). On the other hand, too much wind is also a nuisance: 'My health pretty well, but only wind doth now and then torment me about the fundament extremely' (31 May 1662; 3: 97).

[12] At the end of the month, Pepys sums up his physical health, lamenting his overreliance on laxatives: 'Myself in pretty good health now, after being ill this month for a week together. But cannot yet come to shit well, being so costive, that for this month almost, I have not had a good natural stool; but to this hour am forced to take physic every night, which brings me neither but one stool, and that in the morning as soon as I get up – all the rest of the day very costive' (31 October 1663; 4: 358). This problem seems to have persisted. Four years later Pepys records, 'I find myself to be full of wind, and my anus to be knit together, as it is always with cold' (16 August 1667; 8: 388). Then again, diarrhea can also cause problems, especially in an unfamiliar house: 'in the night was mightily troubled with a looseness (I suppose from some fresh damp Linnen that I put on this night); and feeling for a chamber pott, there was none, I having called the maid up out of her bed, she had forgot I suppose to put one there; so I was forced in this strange house to rise and shit in the Chimny twice; and so to bed and was very well again' (28 September 1665; 6: 244).

[13] Mennes was appointed to Comptroller of the Navy in 1661 – a post he held for ten years. Pepys was frequently insulting about the generation gap between them, referring to Mennes as 'the old foole' (7 April 1663; 4: 98); 'the old coxcomb' (23 May 1663; 4: 152); suggesting that Mennes deserved to 'be pitied for his dotage' (12 March 1664; 5: 80); describing how he went 'to the fleet like a doting fool, to do no good but proclaim himself as asse' (16 September 1665; 4: 226); calling him 'a simple weak man' (17 September 1665; 6: 228) and 'worse then nothing' (27 September 1666; 7: 300). On one occasion Pepys laments 'how ill all the Controller's business is likely to go as long as ever Sir J. Mennes lives' (4 December 1668; 8: 384).

List of Works Cited or Consulted

Aarne, Antti, and Stith Thompson. *The Types of the Folktale: A Classification and Bibliography*. 2nd edn. FF Communications 184. Helsinki: Academia Scientiarum Fennica, 1961.

Académie Royale des Médailles. *Médailles sur les principaux évenements du regne de Louis le Grand avec des explications historiques, par l'Académie Royale des Médailles et des Inscriptions*. Paris: Imprimerie Royale, 1702.

———. *Médailles sur les principaux évenements du regne entier de Louis le Grand avec des explications historiques*. Paris: Imprimerie Royale, 1723.

Antonioli, Roland. *Rabelais et la médecine*. (*Études Rabelaisiennes* 12). Geneva: Droz, 1976.

A Rebours, Luc. [Jannet, Pierre.] *Bibliotheca Scatologica*. Scatopolis: Chez les marchands d'aniterges, l'année scatogène 5850 [Paris: P. Jannet, 1850].

Bachorski, Hans-Jürgen. 'Wie der Narr ins Irrenhaus kommt. Diskursdifferenzierung im 16. und 17. Jahrhundert.' *Das Berliner Modell der Mittleren Deutschen Literatur. Beiträge zur Tagung Kloster Zinna 29.9.–01.10.1997*. Ed. Christiane Caemmerer et al. Chloe 33. Amsterdam/Atlanta: Rodopi, 2000, 273–98.

———, and Werner Röcke. 'Narrendichtung.' *Deutsche Literatur. Eine Sozialgeschichte*. Ed. Horst A. Glaser. 9 vols. Reinbek: Rowohlt, 1991. 2: 203–13.

Bächtold-Stäubli, Hanns. *Handwörterbuch des deutschen Aberglaubens*. Berlin: W. de Gruyter, 1927–42.

Bakhtin, Mikhail M. *The Bakhtin Reader*. Ed. Pam Morris. London: Edward Arnold, 1994.

———. *Rabelais and His World*. Trans. Hélène Iswolsky. Bloomington, IN: Indiana University Press, 1984.

Barker, Francis. *The Tremulous Private Body: Essays on Subjection*. London and New York: Methuen, 1984.

Barnaud, Jacques. *Pierre Viret, sa vie et son œuvre (1511–1571)*. [1911]. Nieuwkoop: B. De Graaf, 1973.

Bartels, Emily C. *Spectacles of Strangeness: Imperialism, Alienation, and Marlowe*. Philadelphia: University of Pennsylvania Press, 1993.

Bayle, Pierre. *Dictionnaire historique et critique*. Paris, 1697.

Bebel, Heinrich. *Facetien: Drei Bücher*. Ed. Gustave Bebermeyer. Leipzig, 1931. Hildesheim: Olms, 1967.

Behrnauer, W.F. *Die Vierzig Veziere oder Weisen Meister*. Leipzig, 1851.

Belletête. *Contes turcs en langue turque, extraits du roman intitulé* Les Quarante Vizirs. Paris, 1812.

Bentzinger, Rudolf. 'Einleitung.' *Die Wahrheit muß ans Licht. Dialoge aus der*

Zeit der Reformation. Ed. Rudolf Bentzinger. Leipzig: Reclam, 1983, 8–40.

Berek, Peter. '*Locrine* Revised, *Selimus*, and Early Responses to *Tamburlaine*.' *Research Opportunities in Renaissance Drama* 23 (1980): 33–54.

Bergson, Henri. *Laughter: An Essay on the Meaning of the Comic*. [1911]. Trans. Cloudesley Brereton and Fred Rothwell. København and Los Angeles: Green Integer, 1999.

Bernstein, Eckhardt. Review of *The End of Conduct: Grobianus and the Renaissance Text of the Subject*, by Barbara Correll. *The Journal of English and Germanic Philology* 98.2 (April 1999): 291–93.

Berrong, Richard. *Rabelais and Bakhtin: Popular Culture in Gargantua and Pantagruel*. Lincoln, NE: University of Nebraska Press, 1986.

Beugnot, Bernard. 'Le corps éloquent.' *Le corps au XVIIe* siècle. Ed. Ronald Tobin. Tübingen: G. Narr, 1995, 17–30.

Beyer, Leo. *Der Nürnberger Stadtteil Mögeldorf: Eine Häusergechichte*. Nuremberg: L. Spindler, 1964.

Bloch, Howard R. *The Scandal of the Fabliaux*. Chicago and London: University of Chicago Press, 1986.

Bobertag, Felix, ed. *Vierhundert Schwänke des sechzehnten Jahrhunderts*. Darmstadt: Wissenschaftliche Buchgesellschaft, 1967.

Boccaccio, Giovanni. *Il decameròn*. Ed. Charles S. Singleton. Bari: G. Laterza, 1955.

Bourke, John G. *Scatalogic Rites of All Nations*. Washington, DC: W.H. Lowdermilk & Co., 1891.

Bowen, Barbara C. 'Il faut donner dedans: sexe et/ou rhétorique dans le *Moyen de parvenir*.' *A French Forum*: *Mélanges de littérature française offerts à Raymond C. et Virginia A. La Charité*. Ed. G. Defaux and J. C. Nash. Paris: Klincksieck, 2000, 107–14.

———. *One Hundred Renaissance Jokes*. Birmingham, AL: Summa Publications, 1988.

———. 'Rabelais et le propos torcheculatif.' *Poétique et Narration*. Ed. François Marotin and Jacques-Philippe Saint-Gerand. Paris: Champion, 1993, 371–80.

Brant, Sebastian. *Das Narrenschiff*. Ed. Manfred Lemmer. 3rd edn. Tübingen: M. Niemeyer, 1986.

Braudel, Ferdinand. *Civilization and Capitalism., 15th–18th Century*. Trans. Sian Reynolds. New York: Harper & Row, 1979.

———. *The Structures of Everyday Life: The Limits of the Possible*. New York: Harper & Row, 1981.

Bremond, Claude. 'L'*Exemplum* médiéval est-il un genre littéraire?: *Exemplum* et littérarité.' *Les Exempla médiévaux*: *nouvelles perspectives*. Ed. Berlioz and Polo de Beaulieu. Paris: Champion, 1998.

Brian, Thomas. *The Pisse-Prophet, or, Certaine Pisse-Pot Lectures*. London, 1637.

Brown, Norman O. *Life Against Death: The Psychoanalytic Meaning of History*. Middletown, CT: Wesleyan University Press, 1950.

Bruscambille. *Les Œuvres de Bruscambille. Contenant ses Fantasies, Imaginations, et Paradoxes et autres discours Comique* [*sic*]. Rouen, 1626.

Bucher, Bernadette. *Icon and Conquest: A Structural Analysis of the Illustrations of De Bry's Great Voyages*. Trans. Basia Miller Gulati. Chicago: Chicago University Press, 1981.

Burnett, Mark Thornton. '*Tamburlaine* and the Body.' *Criticism* 33.1 (Winter 1991): 31–47.

Burton, Jonathan. 'Anglo-Ottoman Relations and the Image of the Turk in *Tamburlaine*.' *Journal of medieval and Early Modern Studies* 30.1 (Winter 2000): 125–56.

Bynum, C.W. 'Bodily Miracles and the Resurrection of the Body in the High Middle Ages' in *Belief in History. Innovative Approaches to European and American Religion*. Ed. T. Kselman. Notre Dame: University of Notre Dame Press, 1991, 68–106.

Calvin, John. *Institution de la religion chrestienne*. Ed. Jean-Daniel Benoît. 5 vols. Paris: Vrin, 1957–63.

———. *Ioannis Calvini Opera quae supersunt omnia*. Ed. G. Baum, et al. 1863. New York: Johnson Reprint, 1964.

———. *Des scandales*. Ed. Olivier Fatio. Geneva: Droz, 1984.

Camporesi, Piero. *The Incorruptible Flesh: Bodily Mutilation and Mortification in Religion and Folklore*. Trans. Tania Croft-Murray. Latin trans. Helen Elsom. Cambridge and New York: Cambridge University Press, 1988.

Carbone, Lodovico. *Cento trenta novelle o facezie. Facezie e Dialogo de la partita soa*. Ed. Gino Ruozzi. Bologna: Commissione per i testi di lingua, 1989.

Carroll, Margaret. 'Peasant Festivity and Political Identity in the Sixteenth Century.' *Art History* 10 (1987): 289–95.

Castiglione, Baldassar. *Il Libro del Cortegiano*. Ed. Amadeo Quondam. Milan: Garzanti, 1981.

Catholy, Eckehard. *Fastnachtspiel*. Stuttgart: Metzler, 1996.

Cave, Terence. *Devotional Poetry in France*. Cambridge: Cambridge University Press, 1969.

Chauvin, Victor. *Bibliographie des ouvrages arabes ou relatifs aux Arabes publiés dans l'Europe chrétienne de 1810 à 1885*. Liège: H. Vaillant-Carmanne, 1892–1922.

Cicero, Marcus Tullius. *De oratore*. Ed. Harris Rackham. Rev. edn. London: W. Heineman; Cambridge, MA: Harvard University Press, 1959–60.

Coakely, Sarah. *Religion and the Body*. Cambridge: Cambridge University Press, 1997.

Correll, Barbara. *The End of Conduct: Grobianus and the Renaissance Text of the Subject*. Ithaca, NY, and London: Cornell University Press, 1996.

Cotgrave, Randle. *A Dictionarie of the French and English Tongues (1611)*. Menston, UK: Scolar Press, 1968.

Cottrell, Robert. *The Grammar of Silence: A reading of Marguerite de Navarre's Poetry*. Washington, DC: Catholic University of America Press, 1986.

The Council of Trent. The Canons and Decrees of the Sacred and Œcumenical

Council of Trent. Ed. and trans. J. Waterworth. London: Dolman, 1848. Hanover Historical Texts Project, 1995. 30 September 2002 http://history.hanover.edu/early/trent.htm.

Crouzet, Denis. *Les Guerriers de Dieu*. Paris: Champ Vallon, 1990.

Cunningham, J.S. Introduction. *Tamburlaine the Great*. Ed. J. S. Cunningham. Manchester: Manchester University Press, 1981.

Dandrey, Patrick. 'Catharsis et mélancholie: l'imaginaire du "corps écrivain" entre l'automne de la Renaissance et le zénith du classicisme.' *Le corps au XVIIe siècle*. Ed. Ronald Tobin. Tübingen: G. Narr, 1995, 31–48.

Davies, Michael. '"Bawdy in Thoughts, precise in Words": Decadence, Divinity and Dissent in the Restoration.' *Romancing Decay: Ideas of Decadence in European Culture*. Ed. Michael St John. Aldershot: Ashgate, 1999, 39–63.

Debaisieux, Martine. 'Tableaux en anamorphose: la Passion du Christ dans la Poésie Baroque.' *Le corps au XVIIe* siècle. Ed. Ronald Tobin. Tübingen: G. Narr, 1995, 129–40.

Debus, Allen G. *The English Paracelsians*. Oldbourne History of Science Library. London: Oldbourne, 1965.

Defaux, Gérard. *Rabelais Agonistes: du rieur au prophète*. (*Études Rabelaisiennes* 32.) Geneva: Droz, 1997.

DeJean, Joan. *The Reinvention of Obscenity : Sex, Lies and Tabloids in Early Modern France*. Chicago and London : University of Chicago Press, 2002.

Demerson, Guy, ed. *Œuvres complètes*. By François Rabelais. Paris: Editions du Seuil, 1973.

Descartes, René. 'Treatise on Man.' *The World and Other Writings*. Trans. and ed. Stephen Gaukroger. Cambridge Texts in the History of Philosophy. Cambridge: Cambridge University Press, 1998.

Désiré, Artus. *Les combatz du fidelle Papiste pelerin romain, contre l'apostat Antipapiste*. Rouen, 1550.

Diepgen, Paul. *Deutsche Volksmedizin. Wissenschaftliche Heilkunde und Kultur*. Stuttgart: Ferdinand Enke, 1935.

Dioscorides, Pedanius Anazerbeus. *De medicinali materia libri quinque...* Parrhisiorium Gymnasio: in off. Henrici Stephani, 1516. McGill Osler Library # folio D 594.

Dollimore, Jonathan. *Radical Tragedy*. Durham, NC: Duke University Press, 1993.

Domenichi, Lodovico. *Facetie, motti, et burle, di diversi signori et persone private*. Venice, 1581.

Douglas, Mary. *Purity and Danger: An Analysis of Concepts of Pollution and Taboo*. New York/Washingon, DC: F.A. Praeger, 1966.

Dubose, Trey. 'Re: Urine.' E-mail to Joseph Tate. 30 October 2000.

du Fail, Noël. *Les Contes et discours d'Eutrapel. Œuvres facétieuses de Noël du Fail*. Ed. J. Assézat. Paris, 1874. Nendeln/Lichtenstein: Kraus Reprint, 1972.

Dundes, Alan. *Life Is Like a Chicken Coop Ladder: A Portrait of German Culture Through Folklore.* Detroit: Wayne State University Press, 1989.

du Pont, Gratien. *Controverses des sexes masculin et feminin.* Toulouse, 1534.

Duval, Edwin. *The Design of Rabelais's* Pantagruel. New Haven, CT: Yale University Press, 1991.

Eberlin, Johann von Günzburg. 'Die fünfzehn Bundesgenossen.' *Ausgewählte Schriften.* Vol. 1 of *Sämtliche Schriften.* Ed. Ludwig Enders. 3 vols. *Flugschriften aus der Reformationszeit* 11, 15, 18. Halle: Max Niemeyer, 1896–1902.

Edwards, Mark U. *Printing, Propaganda, and Martin Luther.* Berkeley: University of California Press, 1994.

Eire, Carlos. *War Against the Idols:The Reformation of Worship from Erasmus to Calvin.* Cambridge: Cambridge University Press, 1986.

Elias, Norbert. *The Civilizing Process. Sociogenetic and Psychogenetic Investigations.* Trans. Edmund Jephcott. Rev. edn. Oxford: Blackwell, 1994.

Eliot, Thomas Stearns. *Collected Poems, 1909–1962.* London: Faber, 1974.

Elwood, Christopher. *The Body Broken: The Calvinist Doctrine of the Eucharist and the Symbolization of Power in Sixteenth-Century France.* New York: Oxford University Press, 1999.

Elyot, Thomas. *The Castel of Helthe.* [1541]. New York.: Scholars' Facsimiles and Reprints, 1937.

Emsley, John. *The 13th Element: The Sordid Tale of Murder, Fire, and Phosphorus.* New York: John Wiley and Sons, 2000.

Erasmus, Desiderius. *The Adages of Erasmus.* Introduction and notes William Barker. Toronto: University of Toronto Press, 2001.

———. *The Colloquies of Erasmus.* Trans. Craig R. Thompson. Chicago: University of Chicago Press, 1965.

———. *Opera omnia.* Amsterdam: North Holland, 1969–.

———. *Praise of Folly.* Trans. Betty Radice. Intro. A.H.T. Levi. Harmondsworth: Penguin, 1993.

Erikson, Erik H. *Young Man Luther: A Study in Psychoanalysis and History.* New York: W.W. Norton. 1958.

Facetiae Facetiarum, hoc est, Joco-seriorum fasciculus. Frankfurt, 1615.

Farley-Hills, David, ed. *Rochester: The Critical Heritage.* London: Routledge and Kegan Paul, 1972.

———. *Rochester's Poetry.* London: Bell and Hyman, 1978.

Fastnachtspiele aus dem fünfzehnten Jahrhundert. Ed. Adelbert von Keller. 3 vols. Bibliothek des litterarischen Vereins in Stuttgart 28–30. Stuttgart: Litterarischer Verein, 1853; *Fastnachtspiele aus dem fünfzehnten Jahrhundert – Nachlese.* Ed. A. v. Keller. Bibliothek des litterarischen Vereins in Stuttgart 46. Stuttgart: Litterarischer Verein, 1858.

Flanigan, C. Clifford. 'Liminality, Carnival, and Social Structure: The Case of Late medieval Biblical Drama.' *Victor Turner and the Construction of Cultural Criticism. Between Literature and Anthropology.* Ed. Kathleen Ashley. Bloomington, IN: Indiana University Press, 1990, 42–85.

Fletcher, John. *The Differences, Cavses, and Ivdgements of Vrine*. London, 1598.

Forssman, Erik, ed. *Marcus Vitruvius Pollio: Zehen Bücher/von der Architectur und kunstlichen Bauen. Erstmals verteuscht durch Gualther Hermenius Rivius*. 1548. Hildesheim and New York: G. Olms, 1973.

Fortescue, Thomas. *The Foreste, or Collection of Histories*. London, 1571.

Foucault, Michel. *Histoire de la folie à l'âge classique: Folie et déraison*. Paris: Plon, 1961.

——. *The History of Sexuality*. Trans. Robert Hurley. New York: Pantheon, 1978.

——. *Madness and Civilization: A History of Insanity in the Age of Reason*. Trans. Richard Howard. New York: Pantheon, 1965.

——. 'A Preface to Transgression.' *Language, Counter-Memory, Practice: Selected Essays and Interviews*. Ed. Donald F. Bouchard. Trans. Donald Bouchard and Sherry Simon. Ithaca, NY: Cornell University Press, 1977, 29–52.

Franck, Sebastian. *Sprichwörter*. Frankfurt am Main: Christian Egenolff, 1541.

Franklin, Alfred. 'Etude sur la voirie et l'hygiène publique à Paris depuis le XIIe siècle' in *Estat, noms et nombre de toutes les rues de Paris en 1636*. Ed. A. Franklin. Paris: Léon Willem, 1873, 1–75.

——. *Paris et les Parisiens au seizième siècle*. Paris: Emile-Paul Frères, 1921.

Freedberg, David. *The Power of Images: Studies in the History and Theory of Response*. Chicago: University of Chicago Press, 1989.

——. *The Prints of Pieter Bruegel the Elder*. Tokyo: Bridgestone Museum of Art. Tokyo: Tokyo Shimbun, 1989.

Freud, Sigmund. *The Standard Edition of the Complete Psychological Works*. Trans. and ed. James Strachey, in collaboration with Anna Freud et al. Vol. XVI, *Introductory Lectures on Psycho-Analysis, III*. London: Hogarth, 1963.

Frey, Jakob. *Gartengesellschaft*. Ed. Johannes Bolte. Tübingen, 1896.

Frischlin, Nicodemus. *Facetiae selectiores...* Strasbourg, 1609.

Fumaroli, Marc, ed. *Fables*. By Jean de La Fontaine. Paris: Livre de Poche, 1985.

Gaignebet, Claude. *A plus hault sens: l'ésotérisme spirituel et charnel de Rabelais*. Paris: Maisonneuve et Larose, 1986.

——, and Marie-Claude Périer. 'L'homme et l'excrétum' in *Histoire des mœurs*. Ed. Jean Poirier. Vol. 1. Paris: Gallimard, 1991, 831–93.

Galen. *Galen's Hygiene [De sanitate tuenda]*. Trans. Robert Green. Springfield, IL: Charles C. Thomas, 1951.

Gast, Johann. *Convivalium sermonum liber*. Basel, 1566.

Gaster, Moses. *The Exempla of the Rabbis*. London: Leipzig, 1870.

Gaudiani, Claire. *The Cabaret Poetry of Théophile de Viau: Texts and Traditions*. Tübingen: G. Narr, 1981.

Gellius, Aulus. *Les Nuits Attiques*. Ed. and trans. René Marache. Paris: Belles Lettres, 1967.

Gesta Romanorum. Ed. H. Oesterley. Berlin: Weidmannsche Buchhandlung, 1872.

Gibson, Walter. *Bruegel*. New York and Toronto: Oxford University Press, 1977.

——. 'Bruegel's Rustic Revels: Laughter in the Dining Room.' Paper presented at the Sixteenth Century Studies Conference in the session 'In Pursuit of Laughter.' Co-chaired by Walter Gibson and Alison Stewart. Toronto: 23 October 1998.

———. *Pieter Bruegel the Elder: Two Studies*. [Lawrence]: Spencer Museum of Art, University of Kansas Press, 1991.

Goethe, Johann Wolfgang von. *Götz von Berlichingen*. Stuttgart: Reclam 1965.

Goldberg, Harriet. *Motif-Index of medieval Spanish Folk Narratives*. medieval and Renaissance Texts and Studies 162. Tempe, AZ: medieval and Renaissance Texts and Studies, 1998.

Greenberg, Mitchell. *Baroque Bodies: Psychoanalysis and the Culture of Absolutism*. Ithaca, NY, and London: Cornell University Press, 2001.

Greene, Graham. *Lord Rochester's Monkey*. New York: Penguin, 1974.

Greimas, Algirdas Julien and Teresa Mary Keane, eds. *Dictionnaire du moyen français*. Paris: Larousse, 1992.

Groß, Angelika. *'La Folie': Wahnsinn und Narrheit im spätmittelalterlichen Text und Bild*. Beiträge zur älteren Literaturgeschichte. Heidelberg: C. Winter, 1990.

Le Guide du patrimoine. Paris: Hachette, 1994.

Guise, Charles de, Cardinal de Lorraine. *L'oraison de Monseigneur le illustrissime & reverendissime Cardinal de Lorraine, faicte en l'Assemblée de Poyssi, le Roy y estant present, le XVI. iour de Septembre. M.D.L.X.I.* Paris, 1561.

Handwörterbuch des deutschen Aberglaubens. Ed. Hanns Bächtold-Stäubli. 10 vols. Berlin/Leipzig: Walter de Gruyter, 1927–42.

Henderson, Jeffrey. *The Maculate Muse. Obscene Language in Attic Comedy*. New Haven and London: Yale University Press, 1975.

Herolt, Johann. *Sermones discipuli de tempore et de sanctis*. Cologne, 1474.

Higman, Francis M. *Piety and the People: Religious Printing in French, 1511–1551*. Aldershot: Scolar Press, 1996.

———. *The Style of John Calvin*. London: Oxford University Press, 1967.

Hippocrates. *The Theory and Practice of Medicine*. Trans. Francis Adams. New York: The Citadel Press, 1964.

Hispanus, Petrus. *Thesaurus Pauperum*. Antwerp, 1476.

The History of the Forty Vezirs or the Story of the Forty Morns and Eves. Written in Turkish by Sheykh-Zada. Ed. and trans. E. J.W.Gibb. London: George Redway, 1886.

Höfler, Max. *Volksmedizin und Aberglauben in Oberbayerns Gegenwart und Vergangenheit*. [1888] Nandeln, Lichtenstein: Sändig Reprint, 1976.

Hollstein, F.W.H. *Dutch and Flemish Engravings Etchings and Woodcuts ca. 1450–1700*. Many vols. Amsterdam: Menno Hertzberger, 1949 ff.

———. *German Engravings Etchings and Woodcuts ca. 1400–1700*. Many vols. Amsterdam: Menno Hertzberger, 1954 ff.

Hope, Geoffrey. 'Tales of Literacy and Authority in the *Violier* (1521): The French *Gesta Romanorum*.' *Bibliothèque d'Humanisme et Renaissance* 59 (1997): 353–63.

Hugo, Victor. *Les misérables*. Ed. Maurice Allem. Paris : Gallimard, 1951.

Hults, Linda. *The Print in the Western World: An Introductory History*. Madison: University of Wisconsin Press, 1996.

The Illustrated Bartsch. Ed. Walter Strauss. Numerous vols. New York: Abaris Books, 1978ff.

Inglis, David. *A Sociological History of the Excretory Experience: Defecatory Manners and Toiletry Technology*. Lewiston: Edwin Mellen, 2001.

Jeanneret, Michel. *Des Mets et des mots: banquets et propos de table à la Renaissance*. Paris: José Corti, 1987.

Jones, Mark. *Medals of the Sun King*. London: Trustees of the British Museum, 1979.

Joubert, Laurent. *Traité du ris, suivi d'un dialogue sur la cacographie française*. 1579. Geneva: Slatkine Reprints, 1973.

Le Journal d'un bourgeois de Paris sous le règne de François Ier (1515–1536). Ed. V.-L. Bourrilly. Paris: Picard, 1910.

'Karsthans.' *Die Wahrheit muß ans Licht. Dialoge aus der Zeit der Reformation*. Ed. Rudolf Bentzinger. Leipzig: Reclam, 1983, 84–126.

Kartschoke, Dieter. 'Narrendichtung.' *Einführung in die deutsche Literatur des 12. bis 16. Jahrhunderts*. Ed. Winfrid Frey et al. 3 vols. Opladen: Westdeutscher Verlag, 1981, 3: 139–64.

Kasten, Ingrid. "'Narrheit' und "Wahnsinn." Michel Foucaults Rezeption von Sebastian Brants "Narrenschiff."' *Festschrift Walter Haug und Burghart Wachinger*. Ed. Johannes Janota et al. Tübingen: Niemeyer, 1992, 1: 233–54.

Kindler, Gerhard. *Mögeldorf einst und jetzt: Ein Bildband*. Nuremberg: Arbeitsgemeinschaft für Belange und Geschichte Mögeldorfs, 1978.

Klibansky, Raymond, Erwin Panofsky, and Fritz Saxl. *Saturn and Melancholy: Studies in the History of Natural Philosophy, Religion, and Art*. New York: Basic Books, 1964.

Könneker, Barbara. *Satire im 16. Jahrhundert: Epoche – Werk – Wirkung*. Munich: Beck, 1991.

Kotin, Armine. *The Narrative Imagination: Comic Tales by Philippe de Vigneulles*. Lexington, KY: University Press of Kentucky, 1977.

Kramer, Heinrich and James Sprenger. *The Malleus Maleficarum of Heinrich Kramer and James Sprenger*. Trans. and intro. Montague Summers. New York: Dover, 1971.

Kramer, Karl-S. *Bauern und Bürger im nachmittelalterlichen Unterfranken. Eine Volkskunde auf Grund archivalischer Quellen*. Würzburg: Komissionsverlag Schöningh, 1957.

Kristeva, Julia. *Powers of Horror: An Essay on Abjection*. Trans. Leon S. Roudiez. New York: Columbia University Press, 1982.

Krumbholz, Eckart. *Martin Luther. 'Euch stoßen, daß es krachen soll.' Sprüche, Aussprüche, Anekdoten*. Berlin: Buchverlag der Morgen, 1983.

Ein kurtzweilig Lesen von Dil Ulenspiegel. Ed. Wolfgang Lindow. Stuttgart: Philipp Reclam, 1978.

La Fontaine, Jean de. *Œuvres de J. de La Fontaine*. Ed. Henri Régnier. Nouvelle édition. Les Grands Écrivains de la France. Paris: Hachette, 1889.

Landau, David, and Peter Parshall. *The Renaissance Print 1470–1550*. New Haven and London: Yale University Press, 1994.

[La Place, Pierre de.] *Commentaires de l'estat de la religion & republique soubs les Rois Henry & François seconds, & Charles neufieme*. N.p., 1565.

Laporte, Dominique. *Histoire de la merde*. Paris: Christian Bourgois, 1993 [1978]. Trans. Rodolphe el-Khoury and Nadia Benabid as *A History of Shit*, Cambridge, MA: MIT Press, 2000.

Lemmer, Manfred. 'Der Erfolg des Werks.' *Sébastien Brant, 500e anniversaire de 'La nef des folz'; 'Das Narren Schyff', zum 500 jährigen Jubiläum des Buches von Sebastian Brant: 1494–1994*. Basel: Christoph Merian Verlag, 1994, 175–81.

Lestringant, Frank. *Une sainte horreur ou le voyage en eucharistie. XVIe–XVIIIe*. Paris: Presses Universitaires de France, 1996.

Lévi-Strauss, Claude. *L'homme nu*. Paris: Plon, 1971.

Levin, Harry. *The Overreacher: A Study of Christopher Marlowe*. Cambridge, MA: Harvard University Press, 1952.

Lewin, Ralph A. *Merde: Excursions in Scientific, Cultural and Socio-Historical Coprology*. New York: Random House, 1999.

Little, Lester K. *Religious Poverty and the Profit Economy in medieval Europe*. Ithaca, NY: Cornell University Press, 1978.

Longhurst, Robyn. *Bodies: Exploring Fluid Boundaries*. London: Routledge, 2001.

Lowe, Peter. *The Whole Course of Chirurgerie*. 1597. Birmingham, AL: Classics of Medicine Library, 1981.

Luther, Martin. *Auff dz ueberchristlich..ueberkünstlich buoch Bocks Emßers zuo Leyptzk. Antwort Martin Luthers. Darinn auch Murrnarrs seines Gesellen Gedocht würt.*. 1521. (Zentralbibliothek Zürich 18.257.20).

———. *Colloquia oder Tischreden Doctor Martini Lutheri...durch Herrn Johann Aurifaber*. Frankfurt am Main, 1567.

Macrobius. *Saturnalia*. Ed. James Willis. Leipzig: B.G. Teubner, 1970.

———. *The Saturnalia*. Trans. Percival Vaughan Davis. New York: Columbia University Press, 1969.

Magdeburg, Mechthild von. *Das fliessende Licht der Gottheit*. Ed. P. Gall Morell. Darmstadt: Wissenschaftliche Buchgesellschaft, 1963.

Magocsi, Paul Robert. *Historical Atlas of East Central Europe*. Seattle and London: University of Washington Press, 1993.

Mainardi, Arlotto. *Motti e facezie del Piovano Arlotto*. Ed. G. Folena. Milan: Ricciardi, 1953.

[Mainardo, Agostino]. *Anatomie de la messe et du messel*. Geneva, 1555.

Mal Lara, Juan de. *Filosofía vulgar*. Barcelona: Selecciones Bibliófilas, 1958–59.

Maple, Eric. *Magic, Medicine & Quackery*. South Brunswick, NJ: A.S. Barnes, 1968.

Marin, Louis. *Portrait of the King*. Trans. Martha Houle. Minneapolis, MN: University of Minnesota Press, 1988.

Marlowe, Christopher. *Tamburlaine the Great*. Ed. J. S. Cunningham. Manchester: Manchester University Press, 1981.

———. *Tamburlaine the Great: Parts I and II*. Ed. Tatiana A. Wolff. London: Methuen and Co. Ltd., 1964.

Martz, Louis. *The Poetry of Meditation*. New Haven, CT: Yale University Press, 1954.

Matheson, Peter. *The Rhetoric of the Reformation*. Edinburgh: T & T Clark, 1998.

Mathieu-Castellani, Gisèle. *La Conversation conteuse: les Nouvelles De Marguerite de Navarre*. Paris: Presses Universitaires de France, 1992.

———. 'Poétique de la scatographie: la 11e nouvelle de l'Heptaméron.' *Poétique et Narration*. Ed. François Marotin and Jacques-Philippe Saint-Gerand. Paris: Champion, 1993, 359–69.

Mayhew, Henry. *London Labour and the London Poor*. Intro. John D. Rosenberg. New York: Dover, 1968.

Megill, Allan. 'The Reception of Foucault by Historians.' *Journal of the History of Ideas* 48 (1987): 117–41.

Menestrier, Claude-François. *Histoire du Roy Louis le Grand par les Medailles, Emblêmes, Devises, Jettons, Inscriptions, Armoiries, et autres Monumens Publics, Recuëillies, et Expliquez par le Pere Claude-Francois Menestrier de la Companie de Jesus*. Paris, 1689.

———. *L'Art des Emblèmes*. Lyon, 1662.

———. *L'Art des Emblèmes, où s'enseigne la morale par le figures de la fable, de l'histoire et de la nature*. Paris, 1684.

Merkel, Johannes. *Form und Funktion der Komik in Nürnberger Fastnachtsspiel*. Freiburg im Breisgau: Klaus Schwarz Verlag, 1971.

Midelfort, H. C. Erik. *A History of Madness in Sixteenth-Century Germany*. Stanford: Stanford University Press, 1999.

Molinet, Jean. *Les Faictz et Dictz de Jean Molinet. Publiés par Noël Dupire*. Vol. 2. Paris: Société des Anciens Textes Français, 1937.

Monestier, Martin. *Histoire et bizarreries sociales des excréments, des origines à nos jours*. Paris: Cherche-midi, 1997.

Montaigne, Michel de. *Œuvres complètes*. Ed. Albert Thibaudet and Maurice Rat. Paris: Gallimard, 1962.

Montanus, Martin. *Gartengesellschaft. Martin Montanus Schwankbücher (1557–1566)*. Ed. Johannes Bolte. Tübingen, 1899.

———. *Wegkürzer. Martin Montanus Schwankbücher (1557–1566)*. Ed. Johannes Bolte. Tübingen, 1899.

Moxey, Keith. *Peasants, Warriors, and Wives: Popular Imagery in the Reformation*. Chicago and London: University of Chicago Press, 1989.

Muchembled, Robert. *L'invention de l'homme moderne: Culture et sensibilités en France du XVe au XVIIIe siècle*. Paris: Fayard, 1988.

Muir, Edward. *Ritual in Early Modern Europe*. Cambridge and New York: Cambridge University Press, 1997.

Murner, Thomas. *Die Narrenbeschwörung*. Ed. M. Spanier. Thomas Murners Deutsche Schriften mit den Holzschnitten der Erstdrucke 2. Berlin/Leipzig: Walter de Gruyter, 1926.

———. *Von dem großen Lutherischen Narren*. Ed. Paul Merker. Thomas Murners

Deutsche Schriften mit den Holzschnitten der Erstdrucke 9. Strasbourg: Karl Trübner, 1918.

Musarum Deliciae: or The Muses Recreation. Comp. John Mennes and James Smith. London, 1655.

Navarre, Marguerite de. *Heptaméron*. Ed. Renja Salminen. Geneva: Droz, 1999.

———. *Le Miroir de l'âme pécheresse*. Ed. Joseph L. Allaire. Munich: Wilhelm Fink Verlag, 1972.

———. *Les Prisons*. Ed. Simone Glasson. Geneva: Droz, 1978.

Nykrog, Per. *Les fabliaux*. Geneva: Droz, 1973.

Orenstein, Nadine, ed. *Pieter Bruegel the Elder: Drawings and Prints*. New York: The Metropolitan Museum of Art. New Haven and London: Yale University Press, 2001.

Ortelius, Abraham. *The Whole Theatre of the World. [Theatrum Orbis Terrarum]*. London, 1606.

Oxford English Dictionary Online. Vers 00.1.1. May 2000. Oxford English Dictionary. 11 May 2000. http://www.oed.com

Paré, Ambroise. *Œuvres complètes*. Ed. J.F. Malgaigne. Vol. 3. Geneva: Slatkine Reprints, 1970.

Parr, Johnstone. *Tamburlaine's Malady and Other Essays on Astrology in Elizabethan Drama*. Westport, CT: Greenwood Press, 1953.

Paster, Gail Kern. *The Body Embarrassed: Drama and the Disciplines of Shame in Early Modern England*. Ithaca, NY: Cornell University Press, 1993.

Pauli, Gustav. *Hans Sebald Beham. Ein kritisches Verzeichnis seiner Kupferstiche, Radierungen und Holzschnitte*. [1901]. Baden-Baden: Valentin Koerner, 1974.

Pauli, Johannes. *Schimpf und Ernst*. Ed. Hermann Österley. Bibliothek des Litterarischen Vereins in Stuttgart 85. Stuttgart: Litterarischer Verein, 1866.

Paulinus, Kristian Frantz. *Neu-Vermehrte Heylsame Dreck-Apotheke Wie nemlich mit Koth und Urin Fast alle/ja auch die schwerste/gifftigste Krankheiten...inn- und äusserlich glücklich curiert worden*. Frankfurt am Main, 1714.

Pepys, Samuel. *The Diary of Samuel Pepys*. Ed. Robert Latham and William Matthews. 11 vols. London: G . Bell and Sons, 1970–83.

Perceau, Louis. *Le Cabinet Secret du Parnasse. Théophile et les libertins*. Paris: Cabinet du Livre, 1935.

Persels, Jeffery C. '"Straitened in the Bowels," or Concerning the Rabelaisian Trope of Defecation.' *Etudes Rabelaisiennes* 31 (1996): 101–12.

———. 'Venereal Verse or, For a Reappraisal of the *contreblason*.' *French Literature Series* 20 (1993): 11–20.

Pessler, Wilhelm, ed. *Handbuch der deutschen Volkskunde*. 3 vols. Potsdam: Akademische Verlagsgesellschaft Athenaion MbH, [1934–38].

Pfeiffer, Gerhard, ed. *Nürnberg – Geschichte einer europäischen Stadt*. Munich: Beck, 1971.

Pinto, Vivian de Sola. 'John Wilmot, Earl of Rochester.' *The Pelican Guide to English Literature*. Ed. Boris Ford. Harmondsworth: Penguin, 1965.

Pirckheimer, Willibald. *Eckius dedolatus*. Ed. Niklas Holzberg. Stuttgart: Reclam, 1983.

Plutarch. *Œuvres morales*. Ed. and trans. Jean Dumortier. Paris: Belles Lettres, 1975.

Poggio, Bracciolini. *Facezie*. Ed. Marcello Ciccuto. Milan: Rizzoli, 1983.

Rabelais, François. *Œuvres complètes*. Ed. Mireille Huchon. Paris: Gallimard, 1994.

———. *The Complete Works of François Rabelais*. Trans. Donald M. Frame. Berkeley: University of California Press, 1991.

———. *Gargantua and Pantagruel*. Trans. Sir Thomas Urquhart and Pierre Le Motteux. Intro. Terence Cave. London: Everyman, 1994.

Racaut, Luc. *Hatred in Print: Catholic Progaganda and Protestant Identity during the French Wars of Religion*. Aldershot: Ashgate, 2002.

Ranke, Kurt et al., eds. *Enzyklopädie des Märchens: Handwörterbuch zur Historischen und Vergleichenden Erzählforschung*. 9 vols continuing. Berlin: Walter de Gruyter, 1975–.

Rawcliffe, Carole. *Medicine and Society in Later medieval England*. 2nd edn. Stroud: Sutton, 1997.

Raylor, Timothy. *Cavaliers, Clubs and Literary Culture: Sir John Mennes, James Smith and the Order of the Fancy*. Newark NJ: Delaware University Press, 1994.

Recorde, Robert. *The Vrinal of Physick*. London, 1547.

Regimen sanitatis salernitanum. Trans. John Harington. New York: Augustus M. Kelley, 1970.

Remshardt, Ralf Erik. 'The Birth of Reason from the Spirit of Carnival: Hans Sachs and *Das Narren-Schneyden*.' *Drama in the Middle Ages. Comparative and Critical Essays*. Ed. Clifford Davidson and John H. Stroupe. New York: AMS Press, 1991, 331–55.

Renaud, Michel. *Pour une lecture du* Moyen de parvenir *de Béroalde de Verville*. Paris: Champion, 1997.

Rigolot, François. 'L'Affaire du "Torchecul": Michel-Ange et l'emblème de la Charité.' *Rabelais en son demi-millénaire*. (*Etudes Rabelaisiennes* XXI. Ed. J. Céard & J.-C. Margolin). Geneva: Droz, 1988: 213–24.

———. 'Leda and the Swan: Rabelais' Parody of Michelangelo.' *Renaissance Quarterly*, XXXVIII 4 (Winter 1985): 688–700.

———. 'Rabelais et la scolastique: une affaire de canards.' *Rabelais's Incomparable Book. Essays on his Art*. Ed. Raymond C. La Charité. Lexington, Kentucky, French Forum Publishers, 1986, 102–23.

Ringsrud, Karen Munson and Jean Jorgenson Linné. *Urinalysis and Body Fluids: A Colortext and Atlas*. St. Louis: Mosby, 1995.

Root-Bernstein, Robert and Michèle. *Honey, Mud, and Other Medical Marvels*. Boston: Houghton Mifflin, 1997.

Roper, Lyndal, *Œdipus and the Devil: Witchcraft, Sexuality and Religion in Early Modern Europe*. London and New York: Routledge, 1994.

Rosebury, Theodor. *Life On Man*. New York: Viking Press, 1969.

Röttinger, Heinrich. *Ergänzungen und Berichtigungen des Sebald Beham-Kataloges Gustav Paulis*. Strasbourg: Heitz, 1927.

Rotunda, D. P. *Motif-Index of the Italian Novella in Prose*. Indiana University Folklore Series 2. Bloomington, IN: Indiana University Press, 1942.

Saba, Guido. *Fortunes et infortunes de Théophile de Viau*. Paris: Klincksieck, 1997.

Sachs, Hans. *The Book of Trades (Ständebuch)*. [Woodcuts by Jost Amman]. Introd. Benjamin A. Rifkin. New York: Dover, 1973.

———. 'Ein faßnacht-spiel mit dreyen personen. Das narren-schneyden.' *Hans Sachs*. Ed. Adelbert von Keller. Bibliothek des litterarischen Vereins in Stuttgart 106. Stuttgart: Litterarischer Verein, 1870, 5: 3–17.

———. *Eygentliche Beschreibung aller Stände auff Erden* [*Book of Trades*]. [Woodcuts by Jost Amman]. Facsimile edition. Zwickau: Ullmann, 1966.

Safty, Essam. 'La décheance physique et ses conséquences dans la poésie baroque en France.' *Dalhousie French Studies* 34 (1996): 17–34.

Sanford, Rhonda Lemke. *Maps and Memory in Early Modern England*. New York: Plagrave, 2002.

Satiren und Pasquille aus der Reformationszeit. Ed. Oskar Schade. 2nd edn. 3 vols. Hannover: Carl Rümpler, 1863.

Schade, Richard. *Studies in Early German Comedy 1500–1650*. Studies in German Literature, Linguistics, and Culture 24. Columbia, SC: Camden House, 1988.

Scheffer, Hans, ed. *Ein therapeutischer 'Thesaurus pauperum' des 14. Jahrhunderts in lateinischen Versen...* Inaugural-Dissertation. Leipzig, 1921.

Schmidt, Josef. 'Luther the Satirist: Strategies and Function of His Satire.' *Proceedings of the Martin Luther Quincentennial Conference...1983*. Ed. Gerhard Dünnhaupt. *Michigan Germanic Studies* 10.1–2 (1984): 32–47.

Schoenfeldt, Michael C. *Bodies and Selves in Early Modern England: Physiology and Inwardness in Spenser, Shakespeare, Herbert, and Milton*. Cambridge: Cambridge University Press, 1999.

Scribner, Robert W. *For the Sake of Simple Folk: Popular Propaganda for the German Reformation*. 2nd edn. Cambridge: Cambridge University Press, 1994.

Seaton, Ethel. 'Marlowe's Map.' *Essays and Studies by Members of the English Association*. Ed. E. K. Chambers. Vol. 10 (1924): 13–35.

Sentence decretale, et condemnatoire au fait de la paillarde Papauté. N.p., 1561.

Seyfarth, Carl. *Aberglaube und Zauberei in der Volksmedizin Sachsens*. Hildesheim: G. Olms, 1979.

Shakespeare, William. *The Complete Works*. Ed. Stanley Wells and Gary Taylor. Oxford: Oxford University Press, 1988.

Shrager, Sidney. *Scatology in Modern Drama*. New York: Irvington, 1982.

Siegel, Rudolph E. *Galen on Psychology, Psychopathology, and Function and Diseases of the Nervous System. An Analysis of His Doctrines, Observations and Experiments*. Basel: S. Karger, 1973.

Sienne, Aldebrandin de. *Le Régime du corps*. Ed. Louis Landouzy and Roger Pépin. Paris: Champion, 1911.

Skelton, R. A. Introduction. Ortelius, Abraham. *The Theatre of the Whole World*. 1606. Amsterdam: Theatrum Orbis Terrarum, 1968.

Snoek, G.J.C. *medieval Piety from Relics to the Eucharist*. Leiden: E.J. Brill, 1995.

Soare, Antoine. 'Cadavres exquis du théâtre baroque.' *Le corps au XVIIe* siècle. Ed. Ronald Tobin. Tübingen: G. Narr, 1995, 119–28.

Sommaire Recueil des signes sacrez, sacrifices, et sacremens instituez de Dieu depuis la creation du monde. Lyons, 1561.

Souiller, Didier. 'D'un art poétique baroque.' *Littératures Classiques* 36 (1999): 45–61.

Spencer, Catherine. 'Promesses du corps textuel: *Dom Juan*, 'cet obscur objet du désir.'' *Le corps au XVIIe* siècle. Ed. Ronald Tobin. Tübingen: G. Narr, 1995, 117–18.

Spiewok, Wolfgang, and Ingmar ten Venne. *Das deutsche Fastnachtspiel. Ursprung, Funktionen, Auffuehrungspraxis*. Greifswald: Reinekeverlag, 1993.

Stallybrass, Peter. 'Footnotes.' *The Body in Parts: Fantasies of Corporeality in Early Modern Europe*. Ed. David Hillman and Carla Mazzio. New York and London: Routledge, 1997, 313–25.

——, and Allon White. *The Politics and Poetics of Transgression*. Ithaca, NY: Cornell University Press, 1986.

Stephens, Walter. *Giants in Those Days*. Lincoln, NE: University of Nebraska Press, 1989.

Stewart, Alison G. *Before Bruegel: Sebald Beham and the Origins of Peasant Festival Imagery*. Aldershot, UK, and Burlington, VT: Ashgate, expected 2005.

——. 'Distaffs and Spindles: Sexual Misbehavior in Sebald Beham's Spinning Bee.' *Saints, Sinners, and Sisters: Gender and Northern Art in medieval and Early Modern Europe*. Eds. Jane L. Carroll and Alison G. Stewart. Aldershot, UK, and Burlington, VT: Ashgate, 2003.

——. 'Large Noses and Changing Meanings in Sixteenth-Century German Prints.' *Print Quarterly* 12 (1995): 343–60.

——. 'Paper Festivals and Popular Entertainment: The Kermis Woodcuts of Sebald Beham in Reformation Nuremberg.' *Sixteenth Century Journal* 24 (1993): 301–50.

——. 'Taverns in Nuremberg Prints at the Time of the German Reformation.' *The World of the Tavern: Public Houses in Early Modern Europe*. Eds. Beat A. Kümin and B. Ann Tlusty. Aldershot, UK, and Burlington, VT: Ashgate, 2002, 95–115.

Sullivan, Margaret. *Bruegel's Peasants: Art and Audience in the Northern Renaissance*. New York: Cambridge University Press, 1994.

Tabarin. *Œuvres complètes de Tabarin*. Ed. Gustave Aventin. 2 vols. Paris: P. Jannet, 1858.

Tabourot, Etienne. *Bigarrures du Seigneur des Accords*. Ed. Guillaume Colletet. 1866. Geneva: Slatkine Reprints, 1969.

Taylor, Larissa. *Soldiers of Christ: Preaching in Late medieval and Reformation France*. New York: Oxford University Press, 1992.

Thomas, Vivien and William Tydeman. *Tamburlaine the Great.* Introduction. *Christopher Marlowe: The Plays and Their Sources.* London and New York: Routledge, 1994, 69–81.

Thompson, Stith. *Motif-Index of Folk Literature.* Rev. ed. Bloomington, IN: Indiana University Press, 1955–58.

Thurn, David H. 'Sights of Power in *Tamburlaine.*' *Critical Essays on Christopher Marlowe.* Ed. Emily C. Bartels. New York: G.K. Hall and Co., 1997.

Tlusty, B. Ann. *Bacchus and Civic Order: The Culture of Drink in Early Modern Germany.* Charlottesville and London: University Press of Virginia, 2001.

Tubach, Frederic C. *Index Exemplorum: A Handbook of medieval Religious Tales.* FF Communications 204. Helsinki: Suomalainen, Tiedeakatemia, 1969.

Verville, Béroalde de. *Le Moyen de parvenir.* Ed. Hélène Moreau and André Tournon. Aix-en-Provence: Publications Université de Provence, 1984.

Viau, Théophile de. *Œuvres poétiques.* Ed. Guido Saba. Paris: Bordas, 1990.

Vigarello, Georges. *Le propre et le sale. L'hygiène du corps depuis le Moyen Age.* Paris: Seuil, 1985.

Vigneulles, Phillipe de. *Les Cent Nouvelles Nouvelles.* Ed. Charles H. Livingston. Geneva: Droz, 1972.

Villeneuve, Arnoul de. *Le Régime tresutile et tresproufitable pour conserver et garder la santé du corps humain.* Ed. Patricia Willet Cummins. Chapel Hill: UNC Department of Romance Languages, 1976.

Le Violier des histoires rommaines. Ed. Geoffrey Hope. Geneva: Droz, 2002.

Viret, Pierre. *Des actes des vrais successeurs de Jésus-Christ et des ses apostres et des apostats de l'Eglise papale.* Geneva, 1554.

———. *Les cauteles et canon de la messe, ensemble la messe du corps de Iesus-Christ.* Lyons, 1563.

———. *Dialogues du désordre qui est à présent au monde, et des causes d'iceluy, et du moyen pour y remédier[...].* Geneva, 1545.

———. *Des principaux poincts qui sont auiourd'huy en different, touchant la saincte Cene de Iesus Christ, & la Messe de l'Eglise Romaine, et de la resolution d'iceux.* N.p., [1566?].

———. *De la vertu, et usage du ministere de la parolle de Dieu, & des Sacremens...* Geneva, 1548.

Vitry, Jacques de. *The Exempla or Illustrative Stories from the* Sermones Vulgares *of Jacques de Vitry.* 1890. Ed. and trans. Thomas Frederick Crane. Publications of the Folk-Lore Society 26. Nendeln/Lichtenstein: Kraus Reprint Ltd., 1967, 98, 230–31.

Wallis, Faith. 'Signs and Senses: Diagnosis and Prognosis in Early medieval Pulse and Urine Texts.' *Social History of Medicine* 13.2 (2000): 265–78.

Weinberg, Florence. 'Rabelais's Isle de Ruach (*Quart Livre* 43–44).' *Romance Languages Annual* (1995): 203–07.

Weiske, Brigitte, ed. *Gesta Romanorum.* Fortuna Vitrea 3, 4. 2 vols. Tübingen: Max Niemeyer Verlag, 1992.

Whetstone, George. *The English Myrror.* London, 1586.

Whorf, Benjamin Lee. 'Science and Linguistics.' *Language, Thought, and Reality.*

Selected Writings of Benjamin Lee Whorf. Ed. John B. Carroll. Cambridge: Technology Press of Massachusetts Institute of Technology; New York: John Wiley and Sons, 1956, 207–19.

Whorton, James C. *Inner Hygiene: Constipation and the Pursuit of Health in Modern Society*. New York: Oxford University Press, 2000.

Williams, David. *Deformed Discourse. The Function of the Monster in Mediaeval Thought and Literature*. Montreal: McGill-Queen's University Press, 1996.

Williams, Gordon. *Dictionary of Sexual Language and Imagery in Shakespearean and Stuart Literature*. 3 vols. London: Athlone, 1994.

Wilmot, John. *The Works of John Wilmot Earl of Rochester*. Ed. Harold Love. Oxford: Oxford University Press, 1999.

Yeats, William Butler. *Yeats's Poems*. Ed. A. Norman Jeffares. London: Macmillan, 1989.

Zarucchi, Jeanne Morgan. '"Ludovicus Heroicus": The Visual and Verbal Iconography of the Medal.' *EMF: Early Modern France* I, 1 (1994): 131–42.

———. 'Medals Catalogs of Louis XIV: Art and Propaganda.' *Source: Notes in the History of Art*, XVII, 4 (1998): 26–34.

Index